JN114760

Yes, you can!
How to eliminate
chronic defects

これならできる！　慢性不良撲滅法

SOUTH-FLOW - The Original and Effective Method of
Quality Improvement Scenario Specializing
in Eliminating Chronic Defects

品質改善シナリオ"サザンフロー"の活用

Kenji Minami　南　賢治
Kiyoshi Katayama　片山　清志
Translated by Kenji Minami

東京図書出版

本文中の「サザンフロー」は日本発条株式会社の登録商標です。

なお、本文中では、®は表記していません。

"SOUTH-FLOW" in this text is a registered trademark of NHK SPRING CO., LTD.

In addition, (R) is not written in the text.

Preface

The phrase "quality first" is often heard not only in the media but also among the manufacturing industry all over the world. The manufacturing industry establishes Quality Management or Quality Control Divisions and assigns representatives to push forward quality improvement activities aimed at defect elimination and conducted by quality improvement project teams and QC circles.

The elimination of chronically occurring "chronic defects" has positive effects on every aspect of the industry, as it improves quality guarantee level and achieves cost cutting. Thus, it is attracting attention from all fields of manufacturing. In this book, the term "elimination of chronic defects" means to phenomenally reduce the incidence of chronic defects to a ratio of ten to one or even down to zero.

Looking at the workplaces which have made improvements, in many cases it was the senior staff and colleagues, that is, fellow workers, who broke the status quo and created a manufacturing site capable of reaching its current high non-defective rate (the yield). In other words, "the breakthrough" requires breaking down what has been built through the valuable labor of fellow workers and senior staff. You can easily understand the psychological complexity of the situation. We can hear the voice of senior staff saying, "We cannot break the status quo with such naive thinking!". However, as restructuring (such as the concentration of people, things, and money) has become more normal than ever, it is an undeniable fact that all the persons tasked with improving their workplaces are under unimaginable psychological friction that is far beyond the imagination of senior staff.

Therefore, in this book, in addition to know-how in efficient quality improvement, we introduce a never-before-seen quality improvement process named SOUTH-FLOW, which shows psychological consideration for business people. For instance, at no point in this process do they need to doubt the colleagues close to them. SOUTH-FLOW is an original quality improvement process from NHK Spring Co., Ltd. with a wide range of applications for quality improvement within entire companies. It was devised by the author, Minami Kenji, while he was a member of the NHK Spring Co., Ltd. SQC Promotion Project Team, and it combines the basics of QC and statistical methods with his experience enacting improvements in various industries. It can be thought of as a guide to achieve "everyone's awareness" by "mieru-ka (visualize)" of the site.

In addition, with the spread of personal computers today, in the spring of 2004, the quality improvement software "JUSE-StatWorks / QCAS (Chronic Defect Elimination version)" equipped with SOUTH-FLOW was put on the market.

We want to devote the precious time we have to finding out the true causes of defects, repeatedly asking ourselves "why?" as we carry out our search based on the Five Gen Principle: "changes at the site, actual products and reality" and "the gap between the ideal and reality." In order to make effective use of a quality improvement scenario that is ready for immediate implementation, it is important for humans to focus on on-site information—leaving the calculations and graphs that can be done on a personal computer to the personal computer. In addition to utilizing a quality improvement scenario, including business reforms to eliminate chronic defects and allocation of the right person to the right place, the utilization of computers is indispensable for modern quality improvement.

In the special quality improvement lecture of the basic course (Tokyo class) of the Union of Japanese Scientists and Engineers (JUSE), we have been introducing "SOUTH-FLOW" as one attempt at "quality control education that will be of immediate use to companies."

We would like to express our sincere gratitude to the following individuals:

Koji Kawamura, the Vice President of NHK Spring Co., Ltd. and Yuichi Nagase, the Company Director, for their direction in unveiling SOUTH-FLOW and encouraging words of "Contribute to society as much as possible!".

Masafumi Mita, the Managing Director of the Business Division of the Union of Japanese Scientists and Engineers, and Ichiro Otsuka, the General

Manager, for listening to the opinion of a graduate of the basic course.

Kenjiro Hayashi, the President of the Institute of Union of Japanese Scientists and Engineers, who accepted my request as a user of their statistical software.

Hikoyasu Shimizu, the Director of Published by JUSE Publishing Co., and Kazuki Fukumoto, the editor of the Publishing Department, for their valuable guidance in publishing.

We also give thanks to:

The improvement leaders of the three companies who provided successful cases of improvement using SOUTH-FLOW.

Mr. Shinichi Rikihisa, a QC consultant who provided valuable advice in creating successful cases.

Mr. Satoshi Hikai, Mr. Yasuyuki Izuhara, Mr. Yasuhiko Ito, Mr. Kiyoji Toda, Mr. Katsuyoshi Kimura, Mr. Masahito Suzuki, Mr. Hideaki Nakanishi, Mr. Kazuaki Suzuki, Mr. Kentaro Ishida, Mr. Yoshiyuki Hirose and Mr. Yoshiyasu Iwaki, for reading the manuscript and giving us useful opinions.

We would like to take this opportunity to thank all concerned parties for their understanding and cooperation.

January 2005

Kenji Minami, Kiyoshi Katayama

Moreover, we also give thanks to:

Ms.Naoko Furukawa, for translating support and Ms.Mayuka Kawaguchi of Sparta English Conversation Co., Ltd., for supporting our translation.

Ms.Kaname Kaneko, Ms.Kaori Usui, Ms.Kana Minami and Mr.Hiroki Minami for reading the manuscript and giving us useful opinions.

We would like to take this opportunity to thank all for their cooperation.

August 2022

Kenji Minami, Kiyoshi Katayama

📖 How to Use This Book

This book is written as a guide to SOUTH-FLOW for the elimination of chronic defects.

(1) The necessity of SOUTH-FLOW when carrying out activities to eliminate chronic defects in a company
(2) SOUTH-FLOW steps / procedures / important points
(3) QC methods useful in SOUTH-FLOW
(4) Successful cases of improvement activities using SOUTH-FLOW
(5) Features of improvement software equipped with SOUTH-FLOW

In addition, this book not only allows individuals to acquire know-how for eradicating chronic defects that can be used as an immediate force in companies, but also supervises and technical staff in the field, including QC circle leaders in group education such as rank-based education. It can also be used as teaching material in educational curriculum for workplace managers.

Figure A shows an example of a quality improvement curriculum utilizing SOUTH-FLOW made using this book and improvement software.

Whether for individual study or as part of training, you can use this book more efficiently if you have made the following preparations.

(1) Understanding of SQC methods such as the QC Seven Tools and the new QC Seven Tools
→ As a tool for visualization, they can be put to use immediately in the workplace.
→ Among the 11 steps of SOUTH-FLOW, Step 7 mentions the L8 orthogonal array experiment as an experiment for factor analysis, but the number of factors that occur may not be large depending on the improvement theme to be tackled, so a solution is also possible using the familiar QC circle way of

"confirmation experiment" as Step 8.

(2) Understanding the QC Story
→ It will be a good benchmark for understanding the characteristics of SOUTH-FLOW.

(3) Awareness of the actual problems, worries, and themes related to eliminating chronic defects
→ Reading with an awareness of the worries and problems, and actively trying to grasp something from the text, will allow you to obtain a much larger amount of information from this book.

Even if you cannot make these preparations, you can utilize them to sharpen your awareness of QC methods and problems while progressing through this book.

Timetable		Curriculum Content	Referenced Chapters and Sections
Day 1	9:00~10:00	The Necessity and Outline of SOUTH-FLOW	Chapter 1 (1.1 to 1.6)
	10:00~12:00	The Steps of SOUTH-FLOW	Chapter 2 (2.1 to 2.11)
	13:00~14:00	SOUTH-FLOW Success Stories	Chapter 4 (4.1 to 4.3)
	14:00~16:00	Improvement Software Training (Using a Tutorial)	Chapter 2 (Improvement Cases) Appendix (Improvement Software)
	16:00~17:00	QC Methods Useful For SOUTH-FLOW	Chapter 3 (3.1 to 3.4)

	17:00~17:15	Day 1 Summary (Review of Important Points)	Chapters 1 to 4
Day 2	8:30~9:00	Psychology and Statistics to Aid Improvement	Coffee Break and Tea Time
	9:00~12:00	Applying Software to Each Person's Improvement Theme	Chapter 2 Appendix (Improvement Software)
	13:00~14:00	Announcement of Improvement Themes (1)	—
	14:00~15:00	Announcement of Improvement Themes (2)	—
	15:15~16:15	Mutual Guidance Practice for Each Improvement Theme	Chapters 1 to 4
	16:15~17:15	Comprehensive Q&A and General Feedback from the Instructor	Chapters 1 to 4

Figure A **Example of a Quality Improvement Training Utilizing SOUTH-FLOW**

【Target Readers of This book】

This book is mainly aimed at managers and supervisors on the front line of quality improvement in their workplace, including those who are in charge of manufacturing, production technology, quality control, quality improvement, and purchasing departments that suffer daily from chronic defects.

We have refined the contents so that QC circle leaders and circle promoters can also use it.

【How to Use This Book】

First of all, we recommend that you read each chapter in order starting from Chapter 1.

(1) Chapter 1 begins with the attitude necessary for an improvement leader to eliminate chronic defects in a company, and proceeds with the know-how for making improvements not alone, but together with your colleagues, while forming the strongest team in your company. We also introduce the points to remember when utilizing SOUTH-FLOW, its necessity, and familiar problems in the workplace.

(2) In Chapter 2, you will gain an understanding of the details of the 11 steps in moving forward with problem solving while conducting efficient fact-based analysis.

You will be placed into a drama-style story based on the actual successes of the leader Mr. Minamiyama as he made improvements using SOUTH-FLOW. From this story, you will be able to see how Mr. Minamiyama, as a problem-solver, got the relevant people involved in making improvements.

(3) Chapter 3 introduces some SQC methods that assist with quality improvement using SOUTH-FLOW.

Most of the methods used in SOUTH-FLOW are also used in QC circles such as the QC Seven Tools, but in this chapter, we will briefly introduce SQC (statistical quality control) methods that are not familiar to QC circles.

We will leave the details of these SQC methods to the SQC technical book.

(4) Chapter 4 contains three cases using SOUTH-FLOW.

You will experience the versatility of SOUTH-FLOW.

In this chapter, you will see how visualizing information is the fastest way to eliminate chronic defects.

(5) Furthermore, as an appendix, in Chapter 2 we will introduce the features of the improvement software used by Mr. Minamiyama.

You can also download the trial version, so please try out this improvement software.

(6) In addition, in the "Coffee Break" section, we introduce some trivia about "human psychology" that will assist with improvement activities using SOUTH-FLOW, and in the "Tea Time" section, we introduce some trivia

about statistics.

Please use these sections for your improvement activities.

Contents

Tea Time

Chapter 1   What is SOUTH-FLOW?

-Quality improvement scenario specializing in eliminating chronic defects-

SOUTH-FLOW is a quality improvement scenario for eliminating chronic defects within a company.

It incorporates know-how on how to find and eliminate the true cause of defects, how to gather the people and information needed, and how to strengthen the improvement team. In addition, this unprecedented quality improvement scenario incorporates know-how on human relationships and how to avoid psychological friction among employees. Chapter 1 introduces SOUTH-FLOW's features for eliminating chronic defects over the course of six sections.

Section 1.1 introduces the most efficient quality improvement scenario, which is formed by combining the SOUTH-FLOW processes of searching for the root cause and forming the strongest team.

Section 1.2 takes a closer look at the two features of SOUTH-FLOW.

Section 1.3 introduces the need for SOUTH-FLOW from the perspective of the manufacturing worksites carrying out activities to eliminate chronic defects.

Section 1.4 touches on the QC methods used in each step of SOUTH-FLOW and introduces what kind of QC method is effective at what point in quality improvement.

Section 1.5 introduces the concerns of people in the position of promoting or practicing quality improvement within their company as well as the answers that SOUTH-FLOW provides, all in a Q&A format.

Section 1.6 introduces the points to keep in mind in order to effectively utilize SOUTH-FLOW in the manufacturing industry.

1.1 What is SOUTH-FLOW?

When eliminating chronic defects, if you do not come across an improvement scenario that leads to efficient improvement, then no matter how hard you try, you may not be able to find a clue as to how to solve your problems. You may give up on your activities, or you may not get the cooperation of those around you. In these cases, your improvement activities become inefficient. There are a number of other challenges in addition to these.

SOUTH-FLOW is an unprecedented improvement scenario in the elimination of chronic defects.

1) The process of searching for the root cause:

Focusing on "lot control" and the "blind spots of lot control" that are the basics of the manufacturing industry, the system is systematized so that the root cause can be found from both searching for factors derived from the result and factors derived from the cause (for example, compared to the traditional methods such as problem-solving QC stories, SOUTH-FLOW refines the "factor analysis" step into six steps to explore the root cause).

2) The process of forming the strongest team:

As the team proceeds with improvement, they can form the strongest team while incorporating key people in the company into the team (for example, at the stage of factor investigation, they form the strongest team by assembling key people in the company into the improvement team).

These are the most efficient quality improvement processes created in order to overcome quality defects.

(1) The Process of Searching for the Root Cause

There was no chronic defect improvement process in the manufacturing industry that could give detailed guidance about what kind of information should actually be collected from the worksite and how to analyze it.

In today's manufacturing industry, when quality improvement is carried out, processes that lead to improvement such as problem-solving QC stories and goal-achieving QC stories have been widely introduced and are used by many companies.

However, the reality is that these processes have to rely on the ideas of those in charge of improvement at each company.

On the other hand, SOUTH-FLOW can be applied to any manufacturing industry due to its specialization as a quality improvement scenario solely for eliminating chronic defects. With SOUTH-FLOW, you can easily see "what kind of data on site should be collected when and which QC methods should be used." Figure 1.1 shows the 11 steps of SOUTH-FLOW and their main substeps.

Figure 1.2 shows a step-by-step comparison between SOUTH-FLOW and conventional problem-solving QC stories. You can clearly see that the investigation corresponding to Step 4 "Factor Analysis" in the conventional problem-solving QC story is subdivided into six steps in SOUTH-FLOW, and the method for guiding factor analysis is significantly different from the traditional problem-solving QC story.

We can find out the true cause of the source and what should be "mieru-ka" and how to analyze it by accumulating the awareness of the people concerned.

Many of the important management points for manufacturing are shared across the industry. A typical example is "lot control," which manages changes in basic manufacturing conditions as lot divisions.

When conducting activities to eliminate chronic defects, we can find clues in various intra-lot and lot-to-lot "differences," moreover in the current "blind spot of lot control," so SOUTH-FLOW conducts research with a multifaceted approach.

To give an example of a "blind spot of lot control," take the manufacturing conditions of materials which were not considered an area of concern in terms of lot control of dimensions directly related to product functions. These could be important changes for appearance defect known as "dent defect."

Each step and procedure of SOUTH-FLOW will be introduced in detail in

Chapter 2 along with improvement cases.

(2) The Process of Forming the Strongest Team

For SOUTH-FLOW, we do not carry out chronic defect elimination activities alone. We devise ways to move forward with improvement while simultaneously bringing in the strong members (key people) needed on our team and strengthening the team through the course of our activities. Figure 1.3 shows the process of SOUTH-FLOW step team formation.

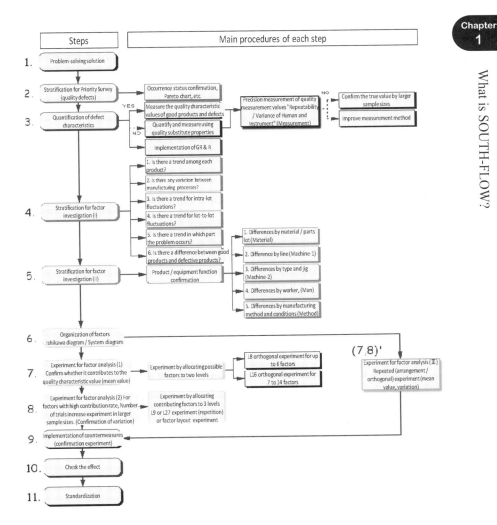

Figure 1.1 SOUTH-FLOW Steps

Steps	QC Story by Problem Solving Methodology	Steps	SOUTH-FLOW
1	"Theme Selection" (1) Select based on what needs to be work on. (2) Select a QC story for the problem you are working on.	1	"Problem-Solving Solution" [Scenario Selections for Pursuing the Root Cause] (1) Select SOUTH-FLOW when defects occur (2) Clarify the reason for selection (based on the necessity to work on), and select the theme (same as column to the left)
2	"Understanding the Current Situation and Setting Goals" (1) Find important problems that have a great influence on the overall problem, (2) Set improvement goals for important problems.	2	"Stratification for Priority Survey" (What to Start With) (1) Find important problems that have a great influence on the overall problem (same as column to the left) (2) Set improvement targets for important problems (same as column to the left) (3) Make an activity plan such as the overall schedule until completion and the division of roles (same as column to the left) (4) Check the measures to prevent the release of defective products to customers.
3	"Creating an Activity Plan" Make an activity plan such as the overall schedule until completion and division of roles		
4	"Factor Analysis" [Ishikawa diagram, etc. [5M]]	3	[Quantification of Defect Characteristics] [Reliability Confirmation of Quality Characteristic Evaluation] (1) Repeatedly evaluate non-defective products and defective products to confirm the reliability of evaluation (measurement). (2) Check the reliability of measurements and if you need members who are familiar with the devices that measure the defects.
		4	"Stratification for Factor Investigation (I) [Phenomenon] [Finding Factors from the Result System] (1) Search for factors from the phenomenon side, such as the occurrence status of similar products and the search for related processes before and after. (2) Select the members necessary to solve the problem from the phenomenon data.
		5	"Stratification for Factor Investigation (II) [Details]" [Search for Factors From the Causative System] While comparing the ideal form of the product and manufacturing process with the reality, the facts of 4M (materials, equipment, people, methods) are verified with data to discover the true cause.

		6	"Organization of Factors" [Organize Factors with the Strongest Members using Facts as Hints] (1) Using the above factor investigation results as a hint, sort out the factors with the team's strongest members who are indispensable for solving the problem. (2) We will dig deeper so that we can experiment with the factors that can be considered in the next step.
		7	"Experiment for Factor Analysis (I)" [Searching for "Contributing factors" From "Possible Factors"] Efficient analysis is performed using orthogonal experiments to search for "contributing factors" from "possible factors".
		8	"Experiment for Factor Analysis (II) [Searching for Optimal Conditions for "Contribution Factors"] (1) Search for the optimal conditions from the "contributing factors" using efficient experiments. (2) If there are few "possible factors," consider them as contributing factors and search for the optimal conditions using efficient experiments.
5	"Countermeasure Planning / Implementation" (1) Develop various measures against the true cause. (2) Make an action plan for the measures and implement the measures.	9	"Implementation of Measures (Confirmation Experiment)" [Introduction of optimal conditions to mass production] (1) Develop measures that address the true cause and can be introduced into mass production (same as column to the left). (2) Make an implementation plan for countermeasures and then implement them (same as column to the left).
6	"Effect Confirmation" (1) After implementing the measures, check the data to see how your results compare to the goal (tangible). (2) Check how much the people involved have grown through the activity (intangible).	10	"Effect Confirmation" [Confirmation of Effects after Introduction of Mass Production Measures] (1) After implementing the measures, check the data to see how your results compare to the goal (same as column to the left). (2) Check how much the people involved have grown through the activities (same as the column to the left).
7	"Standardization and Establishment of Management" Standardize the procedures and management methods necessary to continuously implement effective countermeasures, and confirm whether the effects are continuously obtained.	11	"Standardization" [Succession of improvement measures to the next generation] Standardize the procedures and management methods required to continuously implement effective countermeasures, and confirm whether the effects are continuously obtained (same as the column to the left).

Figure 1.2 **Step-by-step Comparison of Problem-Solving QC stories and SOUTH-FLOW**

SOUTH-FLOW Step No.	The process of forming the strongest team	Description of process time
Step 1	I . Leader's determination	It's time for improvement promoters to decide to believe in solving the problem.
Step 2	II . Uplifting team mind	It is time to aim to create a good mood for improvement with the cooperation of our core colleagues.
Step 2 · Step 3	III. Incorporation of key people	When improving a certain chronic defect, it is time to assemble the strongest members and proceed with the improvement even if the measurement method, other products, and other process managers are not in the initial improvement members. b
Step 4 | Step 10	IV. Improving the solution	It is time for the strongest members to hone their "eyes to discover differences" by visualizing on-site information.
Step 11	V . Further growth Other improvement teams — Other improvement teams	After investigating the cause of chronic failure, the activity presentation that discovered the root cause and improved it impresses many people. It's time to deepen your confidence by feeling how you have impressed your audience. It is also a time for further growth by listening to other presentations.

⬤ : Improvement promoter himself (improvement leader)

⬤ : Companions who are the core of improvement (motivated, positive)

◯ : Other improvement team members

◯ : Key people other than improvement team members

⬤ ⬤ ◯ ◯ : The strongest team members with improved resolution

Figure 1.3 SOUTH-FLOW Steps and the Process of Forming the "Strongest Team"

As shown in Figure 1.3, SOUTH-FLOW begins with the "leader's determination" of the improvement leader in Step 1, "uplifting the team mind" in Step 2, "incorporating key people" in Steps 3-4, "improved resolution" in Steps 4-10, and "further growth" in Step 11. We will promote activities that make the most of the team power while forming the strongest team through these five periods.

The process of forming the strongest team is an important process in promoting improvement along the SOUTH-FLOW, so the five periods are introduced as below.

I. Leader's Determination (Step 1)

I. At the time of the leader's determination, as in Step 1 "Problem-Solving Solution" of SOUTH-FLOW, clarify the theme name and the reason for selection, and register the improvement members.

The first thing I would like you to understand during this period is the role of the improvement promoter himself, the so-called improvement leader. When eliminating chronic defects, improvement promoters believe that they can always solve the problem and carry out improvement activities.

As it is said that "a company is a person," not limited to the manufacturing industry, the problems cannot be improved unless the employees have the vitality to improve it.

The results are even more obvious if the problem is a long-term chronic defect in the process.

Improvement leaders should use SOUTH-FLOW to decide in their own mind to eliminate the themed chronic defects.

II. Uplifting Team Morale (Step 2)

II. At the time of raising the team morale, that is, the time of creating a mood for improvement as an improvement team, Step 2 of SOUTH-FLOW "Stratification for Priority Survey" is performed. In this step, you will check the occurrence of chronic defects and set goals. Furthermore, we will confirm that measures are being taken to prevent the chronic defects listed in the improvement theme from leaking to customers.

For improvement leaders who have decided to eliminate chronic defects at the time of "I. Leader's Determination," the next thing they need is a companion who will be the core of improvement. It is necessary to always create a good improvement mood regardless of the improvement theme. It is effective to know the "2-6-2 Principle" that is often said to raise team morale. Applying this "2-6-2 Principle" to the quality improvement team, it is considered that the following groups will be formed in a group of a certain number of people.

They are called:
1) Group 1 (20%): Ambitious and supportive positives
2) Second group (60%): Centrists depending on the atmosphere
3) Third group (20%): Perverse person? Non-cooperative negatives

This is one of the promotion tips that improvement promoters should never forget in raising the team mind. I think that there are many people who naturally learn and practice even if they have never been aware of it as a "principle." Figure 1.4 shows the atmosphere of improvement based on the "2-6-2 Principle."

As shown in Figure 1.4 (a), when 80% of the people (1st and 2nd groups) are supportive of improvement, you will find it is wonderful that the workplace itself is full of vitality for improvement. On the contrary, the improvement leader must be careful not to create the atmosphere shown in Figure 1.4 (b).

This "2-6-2 Principle" is an important point, as it shows how to create the atmosphere as shown in Figure 1.4 (a), I will touch on it in Section 1.6 (4) again.

III. Incorporation of Key People (Steps 3 and 4)

III. At the time of taking in a key person, Step 3 "Quantification of Defect characteristics [Reliability Confirmation of Quality characteristic Evaluation]" and Step 4 "Stratification for Factor Investigation (I) 【Phenomenon Aspect】 [Factors from Results-Derived Defects]" is performed.

In this step, we check the evaluation (measurement) accuracy of

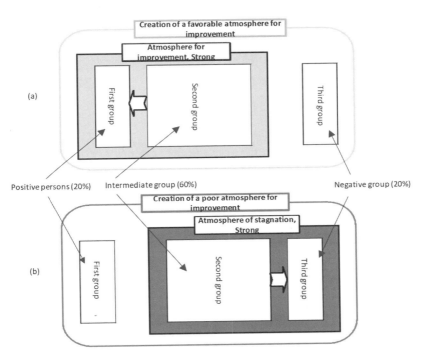

Figure 1.4 Improvement Atmosphere Based on the "2-6-2 Principle"

gauge R & R, etc., and search for factors from the phenomenon side derived from the result such as the relationship between similar products and pre- and post-processes.

It's time to ask other necessary members to join the improvement team if these findings reveal that there are key players outside of the team.

The next thing is that the improvement team, who was able to create a good improvement mood, to confirm whether their essential unique techniques are sufficient in order to eliminate chronic defects and to introduce a new key person if necessary.

SOUTH-FLOW confirms the need for key players in the following fields at this time.

    i) Field of measurement development
    ii) Field in charge of similar products
    iii) Field in charge of other processes

The survey order is such that the initial improvement members can participate in the improvement as soon as necessary. When improving chronic defects in the workplace, it is rare that the people in this measurement, other products, and other processes are involved in the members of the improvement team from the beginning.

The following is a list of key person discovery cases across departments.

i) Fields of measurement development

- A case in which the required accuracy is high in the product specifications of the customer, but the measurement method decided at the development stage remains the same as before, so that an appropriate evaluation cannot be made and a chronic defect occurs.
- A case in which the required accuracy by the customer has not changed, but the evaluation accuracy has declined due to a change in the product shape of the new product, resulting in chronic defects.

ii) Field in charge of similar products

- Similar defects have occurred in similar products before, but the countermeasures had already been taken and no information was provided to the designers or production engineers in charge of other products, resulting in chronic defects.
- The product itself is similar, but the parts purchased are different, and one of the parts purchased is chronically defective because no one noticed that there were errors in the method of taking manufacturing standards.

iii) Field in charge of other processes

- Cases in which the product was chronically defective in their manufacturing process, but was chronically defective without noticing that it was a quality characteristic that was affected by the performance of the previous process.
- Since there is an important process in manufacturing a certain product, they have been required to improve, but they have become chronically defective without noticing that the quality characteristics are greatly influenced by the post-process.

These are just a few of the many cases, but SOUTH-FLOW asks if improvement leaders and their team members need new key people during the "III. Incorporating Key People" period.

If members, including improvement leaders, cannot get cooperation, even if they convey the survey results to key people in other departments, do not give up. Consult with their boss to break the boundaries between departments and sections so that they can carry out activities beyond that (please refer to the improvement cases in Chapters 2 and 4).

IV. Improving the Solution (Steps 4 to 10)

When "III. Incorporating Key People" is completed and the strongest members in the company can be gathered, the next time is "IV. Improving the Solution."

During this period, SOUTH-FLOW identifies factors by the two steps below:

Step 4 "Stratification for Factor Investigation (I) [Phenomenon] [Finding Factors From the Result System]" and

Step 5 "Stratification for Factor Investigation (II) [Search for Factors From the Causative System]",

and Step 6 "Factor Organization" will be performed.

From Step 7 to 10 are "Factor Analysis" to "Measure Planning / Implementation / Effect Confirmation." They are the highlights to show the strongest team member's analytical abilities by their unique technology and visualization capabilities to others.

When making improvements, it is important to always be aware of the problem and improve your eyes to discover differences. During this period, various multifaceted factor investigations will improve the eyes that can tell the difference.

It is also essential to improve your memory during this period. Many people may wonder why improving memory is necessary for quality improvement activity. Chronic defects that occur in the process are often unexpectedly and unknowingly reflected in the minds of people in the workplace. In other words, even if you find something new, you tend to overlook it without thinking that you have found it. People are better at unraveling from associative and chained memories of some hints than starting to write ideals on a white board. In addition, memory is maintained for a long time by repetition.

It is indispensable to connect repetition and chain memory to problem awareness through positive improvement activities. As everyone who is interested in improvement and can always improve quality defects, SOUTH-FLOW gains a lot in the investigation. This is a quality improvement scenario that leads to improvement activities where the awareness of hints inspires people in the workplace.

V. Further Growth (Step 11)

It's time for "V. Further Growth" as the next step after the chronic defects can be eliminated brilliantly through "IV. Improving the Solution." At this time, always keep in mind the active standardization and report on the activities as Step 11 of SOUTH-FLOW is "Standardization [Succession of Improvement Measures to the Next Generation]".

At the final stage, it is recommended that members who have been able to improve chronic defects try to announce improvements inside or outside the department before entering the next improvement theme.

Elimination of chronic defects by SOUTH-FLOW is to pursue and solve the root cause by repeating why the current problem is not in the "state that should be."

When an improvement is announced, the listener is impressed by the fact that the root cause of the occurrence has been successfully identified and resolved. Depending on the idea, it is easy to imagine that there will be a big difference in emotions compared to the announcement that the root cause was unclear and could be improved. In addition, the members who announced will feel the excitement of the audience.

In addition, members of the improvement team who practiced SOUTH-FLOW will be able to imagine how they can work by opening other improvement announcements. At the same time, they will experience a sense of accomplishment and satisfaction in their improvement, and will achieve further growth.

1.2 Features of SOUTH-FLOW

SOUTH-FLOW has the following features that are incorporated into the two processes of "searching for the root cause" and "forming the strongest team" introduced in the previous section.

1) You can collect "people and information" necessary for organizing factors.
2) This is a survey with little "analysis loss" and "psychological friction."

These two characteristics are rarely seen in conventional problem-solving methods. We will introduce these features in detail.

(1) You can correct the necessary "people and information" for organizing factors.

In SOUTH-FLOW for the elimination of chronic defects, linguistic data is organized in Step 6 as a gathering of the wisdom of related parties such as Ishi-kawa diagrams and system diagrams at the stage of factor organization.

As those who have experienced factor sorting methods such as Ishikawa di-agrams know, these methods largely depend on the unique technical capabilities of the people who have come together, and the time required to create them is not short.

Therefore, in order to use these methods efficiently when eliminating chronic defects, the following two points are important.

i) Gather people with a lot of unique technical abilities,

ii) Show useful information on many sites for consideration.

Therefore, in order to deal with these issues, SOUTH-FLOW will investi-gate the current situation of Step 4-1 "Investigation of Defect occurrence Status by Product" and Step 4-2 "Investigation of Fluctuation Status Between Process-es" for the issue i). By doing so, check if the required key person is an original improvement member.

If there are not enough key people needed to eliminate the chronic defect, it is needed to assemble key people within the company based on the results of these surveys.

For issue ii), after incorporating the key people in Steps 4-1 and 2, the fac-tor search survey from the result system in Steps 4-3 to 5, searching for factors derived from the cause in Steps 5-1 to 5 are performed. Then visualize various multifaceted survey results.

Therefore, in SOUTH-FLOW, key people can be assembled as needed, and the strongest members for elimination come together while using multifaceted information on the site as hints will sort out the factors and pursue the true cause.

For the above reasons, in the activities to eliminate chronic defects by SOUTH-FLOW, it is taboo to make a Ishikawa diagram only with the initial improvement members at the initial stage of improvement.

(2) This is a survey with little "analysis loss" and "psychological friction".

In SOUTH-FLOW, among the main control items of quality control, 5M (man [Man], material [Material], machine [Machine], method [Method], measurement [Measurement]), when making improvements within the company. This describes the investigation procedure that leads to "loss in analysis" and psychological friction between employees.

The items to be specified below and the reasons for them are introduced below.

i) Measurement [Measurement] should be investigated in Step 3 of the initial stage of improvement.

ii) People / Workers [Man] should be investigated in Step 5-4 in the latter half of the factor investigation.

iii) Method [Method] should be investigated in th Step 5-5 of the factor investigation.

In SOUTH-FLOW, regarding i) the measurement survey, it is confirmed at the initial stage of improvement after performing Step 2 "Stratification for Priority Survey."

This is because the utilization of the QC method is a prerequisite for data reliability, and if all surveys are wasted because the individual survey data is unreliable after visualizing and analyzing the survey data at the site, it will be a big loss in terms of analysis.

On the other hand, we will investigate the penultimate one in the latter half of the factor investigation regarding i) the investigation of people and workers.

This is because on-site cooperation is indispensable for eliminating chronic defects activities, and it is not effective to conduct a suspicious investigation of workers who should be key people of the improvement team from the beginning of improvement.

Who would be happy to be asked, "What do you think is the cause of the defect in your work?" from the beginning of the chronic defect elimination activity?

Furthermore, iii) method, manufacturing conditions, and construction method will be confirmed at the end of the factor investigation. This is because the staff and staff who are familiar with the manufacturing technology of the process to be improved, like the workers, are the key people of the improvement team. It is meaningless to investigate the manufacturing technology staff from the beginning of the improvement.

From the beginning of the chronic defect elimination activity, if you are asked, "I think the cause of the defect is your set manufacturing method or manufacturing conditions are not optimal. How about manufacturing?". Technicians often do not want to work with this improvement leader.

When aiming to break through the current situation of eliminating chronic defects in various departments within the company, consider both "loss in analysis" and "loss due to psychological friction" promotes efficient improvement within the company.

1.3 Necessity of SOUTH-FLOW

This is the introduction of the necessity of SOUTH-FLOW, which can be seen from the improvement case presentation.

In eliminating chronic defects, it can be said that "improvement activities that transcend the boundaries of departments and staff" are extremely powerful. Furthermore, when thinking about Japanese companies, in order to eliminate chronic defects, it is better to think clearly about the organization and the individual to promote quality improvement.

In many Japanese manufacturing industries, quality defects that can be solved by the staff are often solved by quality improvement activities such as small group activities at the worksite. In addition, looking back on the activities that have won top prizes in quality improvement announcements such as QC circles inside and outside the company, there are many cases in which deep-root-ed root causes have been discovered and resolved through cross-functional improvement activities that transcend the boundaries of the department.

Thinking about why we were able to carry out cross-functional activities that could win top prizes, we would like to point out the following:

- First, the enthusiasm and ingenuity of the improvement team
- Second, strong support led by the boss
- Third, a company with a culture that allows free interaction with other departments and sections

On the other hand, regardless of labels decided by the organization, there are people who are trusted by others as problem solvers. They will say, "That person will solve it again."

Such problem-solvers manage to solve a given chronic defect improvement theme, even if they don't have the specific skills of that theme.

Are the people who can deal with any difficult problem, as they have answers to any questions, special? In addition, SQC teachers outside the company can be said to be the ultimate problem solver because they can give advice to solve the poor dominance of the company. Here, you will notice that the problem solver inside the company and the SQC teacher outside the company have something in common.

Even if they do not have their own skills, they are both masters of collecting a lot of useful information about solving chronic defects, and by visualizing that information, they can draw ideas from people who are familiar with the subject.

After all, when aiming to eliminate chronic defects, the improvement leader himself cannot choose strong support led by his boss or organizational culture.

Therefore, if there is enthusiasm for improvement leaders, there may be a need for a quality improvement scenario that can guide activities to eliminate chronic defects even if the person does not have unique skills.

On the other hand, although one has learned the SQC method, there is also a need for a quality improvement scenario that solves the problem of not being able to practice it when it comes to actually eliminating chronic defects in-house. If there is a quality improvement scenario that can assemble a key person to solve the defect in the activity, it must be possible to eliminate the chronic defect brilliantly.

Therefore, in order to nurture people who can eliminate chronic defects, we considered the following points of view.

- First, changing the corporate culture and organization is not easy and takes time.
- Second, if you teach motivated individuals the two things,

    1, the process of searching for the root cause and

    2, the process of forming the strongest team to eliminate chronic defects,

    they will grow to be a problem-solver that eliminate defects with or without unique skills.
- Third, as the number of problem-solvers in the company increases, many colleagues will be inspired and a culture of improvement that transcends the organization will naturally be created.

Therefore, SOUTH-FLOW, which was conceived to eliminate chronic defects in the manufacturing industry, is a quality improvement process in which cross-functional improvement activities can be naturally carried out while assembling key people for improvement.

SOUTH-FLOW systematizes the method of convening key people that transcends organizational boundaries, and it is shown at Step 3 "Quantification of Defect Characteristics [Reliability Confirmation of Quality Characteristic Evaluation]" and Step 4 "Stratification for Factor Investigation (I) [Phenomenon] [Finding Factors From the Result System] as in Figure 1.1 of the above.

--- **Tea Time** ---

(1) Be a solver rather than an analyst

In companies, QC training inside and outside the company is conducted in various ways.

Most of the training is related to statistical analysis methods (analysis tools) such as the QC Seven Tools, multivariate analysis, and design of experiments.

- Difference Between Training and Field Practice -

Did you know that there are some prerequisites for using analysis tools in the training as follows?

1) The reliability of the data value itself is high.

2) Samples obtained from each population should be representative of the population.

Also, the most difficult thing to teach in these training sessions is the technique of "what data can be collected and how can be improved efficiently in the workplace."

In the quality-related training so far, even if it was an external specialized training, although it was possible to perform statistical analysis as a so-called analyst, it is extremely difficult to acquire the applied ability to solve various themes of the company as a solution desired by the company.

Therefore, students from companies must make improvements while understanding the gap between the training of 1) and 2) above and the reality.

- Training with People Who Can Solve Problems -

Those who can actually solve problems have the ability even if they are not familiar with statistics before receiving the training. Conversely, those who have no practical experience of problem solving have difficulty solving problems even after the training.

Training to become a solver will require attempts never tried before. Isn't training that improves the solution ability to discover the root cause

> while making full use of visualization methods in quality improvement
> what is needed today?

1.4 QC Methods Used in Each Step of SOUTH-FLOW

SOUTH-FLOW recommends effective QC methods for improvement using the QC Seven Tools at each step. Figure 1.5 shows the QC methods used at each step of SOUTH-FLOW. These methods are all effective for visualizing (mie-ru-ka) on-site information.

We introduce the QC methods that are particularly effective in each step from priority survey to factor resolution.

1) An Effective QC Method in Step 2: Pareto Charts

- Visualize the frequency of appearance and cumulative sum on a Pareto chart for each stratified factor and type of characteristics, and confirm what is the most efficient way to improve.

2) An Effective QC Method in Step 3: Gauge R & R (GR & R)

- After deciding the quality characteristics to be improved, if the characteristics can be measured, check whether the measurement accuracy is sufficient for the product standard by GR & R (see Chapter 3 for details).

3) An Effective QC Method in Step 4: Control Chart

- For the change points between lots with different manufacturing conditions, plot the data in chronological order on the control chart and visually check the trend etc.

4) Effective QC methods in Step 5: Histogram, Weibull Analysis

- When the same product is produced on multiple lines, the distribution of data is visualized using a histogram to see the differences between the lines.
- Visualize by Weibull analysis when you want to check whether wear or durability problems due to different materials of molds and jigs are the cause of chronic defects.

5) An Effective QC Method in Step 6: System Diagram

- Repeatedly ask why using a system diagram for the language data obtained based on the "hints" and "experiences" confirmed in the scenarios so far, and sort out possible factors.

6) An Effective QC Method in Step 7: L$_8$ Orthogonal Array Experiment

- When there are many possible factors (up to 6),an L$_8$ orthogonal array experiment helps you to confirm the entangled factors and influence on the quality characteristics by the orthogonal array experiment.
- If there are up to two possible factors, perform a factor layout experiment in the next Step 8 (see Chapter 3 for details).

7) An Effective QC method in Step 8: Factor Placement Experiment

- Perform factor layout experiments to find optimal conditions for factors that contribute to quality characteristics (see Chapter 3 for details).

The above are general QC methods recommended for SOUTH-FLOW. We think there are many methods that are also used in small group activities such as QC circles.

We want to emphasize here that GR & R used in Step 3 is able to clarify the ratio of measurement error with respect to product standards only from the calculated data by statistical software which measured by three measurers 10 to 30 same samples for three days.

For the Weibull analysis in Step 5, it is also a good idea to check the presence or absence of factors due to the wear system of the tool by photographs, etc., as shown in Figure 2.32 of the improvement example in Chapter 2.

In addition, it is advisable to utilize the L$_8$ orthogonal array experiment in Step 7 by getting advice from a person who has been taught in-house as a keyperson in the analysis.

For example, the L$_8$ orthogonal array experiment is one of the most efficient methods in which the result can be obtained by eight experiments where 64 experiments should be performed. In addition, if there are few possible factors or contributing factors, it can be solved by the "confirmation experiment" that is familiar to QC circles.

Steps of SOUTH-FLOW	Ishikawa diagram	Pareto chart	Graphs	Check sheet	Histogram	Scatter plot	Control chart	The New QC Seven Tools (N7)	Correlation analysis	Multivariate analysis	DOE (factor layout experiment)	DOE (orthogonal array experiment)	Reliability engineering	Multivariate association diagram	Gauge R & R (GR&R)	Others (Photo, Video)	Remarks
											※	※	※	※	※		* Explained in Chapter 3
1. Problem-solving solution [Scenario Selections for Pursuing the Root Cause]			◎														Graphs (radar charts, pie charts, bar graphs)
2. Stratification for priority survey (What to start with)		◎	○				○										Pie charts, N7(matrix diagram)
3. Quantification of defect characteristics [Reliability confirmation of quality characterization]			○			○			○						◎		Graphs (line graph, bar graphs)
4. Stratification for factor investigation (I) [phenomenon aspect]		◎	◎	◎	◎	○	◎	○	◎					◎		◎	Graphs (line graph), N7(matrix diagram), Others (photos, videos)
5. Stratification for factor investigation (II) [Detailed aspects]			○	○	◎		◎		○				◎	◎		◎	Graphs (line graph, bar graphs) Others (photos, videos)
6. Organizing of factors [Facts as hints to sort out factors with the strongest members]	◎							◎									N7(system diagram)
7. Experiment for factor analysis (I) [Searching for "contribution factors" from "possible factors"]									○		◎						N7(matrix diagram)
8. Experiment for factor analysis (II) [Searching for optimal conditions for "contribution factors"]									○		◎	◎					N7(matrix diagram)
9. Implementation of measures (confirmation experiment) [Introduction of optimal conditions to mass production]									○								N7(matrix diagram)
10 Effect confirmation [Confirmation of effects after introduction of mass production measures]		◎	◎		◎		◎										Graphs (line graph, pie charts, bar graphs, band graph, radar charts) Others (photos, videos)
11. Standardization [Passing on improvement measures to the next generation]			○	○			◎									○	Graphs (line graph) Others (photos, videos)

Note) ◎ : Especially recommended QC methods, ○ : Recommended QC methods

Figure 1.5 QC Methods Used at Each Step of SOUTH-FLOW

In addition to the learning method of learning the method first and then using it for quality improvement, if you know what kind of method is effective when checking "when and what" in the field like this book, you will be able to learn a lot differently than before.

1.5 Q&A on Utilizing SOUTH-FLOW

Up to this point, we have had the opportunity to hear the concerns of many people from both inside and outside the company. Here, we will introduce problems that are likely to be encountered in quality improvement, especially those from the manufacturing industry, as well as the advantages of using SOUTH-FLOW to resolve them, and the advantages of using SOUTH-FLOW-equipped improvement software.

In addition, the frequency with which we hear these problems is indicated by the [★] mark, and the more [★] marks there are (up to a maximum of 5 ★), the more people are currently troubled by the problem.

Q1 **I can't use the QC methods I learned in Q1 training effectively. [★★★★★]**

I have learned QC methods, but I cannot use them to improve quality within my company. In the end, I only use KKDH (intuition, experience, courage, and completeness) to improve quality.

A1

SOUTH-FLOW guides the improvement team in eliminating chronic defects, showing concretely what type of data to collect in the field, and which QC methods should be used (which are effective). Following the steps of SOUTH-FLOW allows for natural visualization with QC methods.

Q2 **I am not confident that I will make a chance discovery by improving chronic defects. [★★★]**

My senior colleague said, "When I was at a loss and about to give up on improving chronic defects, a manufacturing colleague of mine (who joined the company at the same time as I) who handled similar products taught me how to devise my own good manufacturing process. In just that one conversation we had while standing, my improvement took a giant jump forward. Without these chance discoveries, it's impossible to improve chronic defects." I am not confident that I will be able to make such a chance discovery.

A2

For SOUTH-FLOW, we move forward with improvement while searching for and assembling key people involved with the necessary products from other companies. In particular, the investigation of similar products is carried out in the first part of Step 4's "factor investigation," so should you be troubled by horizontal deployment omission within your company, you can solve it at the initial stage of improvement. Therefore, when you arrive at the end of your improvement efforts, you will certainly not be saying to yourself "it was all due to chance…" as in the case of Q2.

Q3 I am having trouble with a lack of responsibility at other manufacturing sites. [★★★★]

The manufacturing process of our product is long, and there are many themes that cannot be solved by improvements within the section of the company I am in charge of. Is there any good way to make improvements jointly with other sections?

A3

For SOUTH-FLOW, we move forward with improvement while searching for and gathering key people involved in the necessary processes. In particular, investigation into which processes are related to the relevant defects is carried out in the second part of Step 4's "factor investigation," so even if you don't have anyone on your improvement team who is in charge of the process that needs to be improved, you can bring in the necessary key people involved in said process.

Q4 I don't know how to tackle a theme with a low defect inci-dence rate. [★★★] 1

The incidence rate is low, but if a defective product is released to a customer, it will cause them a lot of trouble. Currently, we are trying to prevent this by double sorting. The incidence of defects is low and the true cause cannot be found.

A4

In SOUTH-FLOW, the key people involved in resolving the relevant defects work together with the strongest members to reach a resolution. Even for themes with a low incidence of defects, upon investigating various factors with the strongest members, we uncover the true cause by employing such techniques as attempting to recreate the relevant defects.

Q5 They won't accept an improvement method that was pre-viously successful. [★★★]

We have solved quality problems due to differences between workers before. When making similar improvements at other factories, we requested that the workers cooperate in collecting data, but neither those in charge of manufacturing nor the workers at the factory cooperated at all.

A5

For SOUTH-FLOW, in our investigation of the phenomenon, we investigate the factors from the results. At that time, the strongest members are called. Next, in our investigation into the details, we investigate the factors from the causes. However, it isn't until the end of the improvement process that the strongest members investigate the responsibility of the workers as well as the current manufacturing methods and conditions. This minimizes the psychological friction of the improvement team members.

Q6 **The Ishikawa diagram does not contribute to improve-ments despite the time required to create it. [★★★★★]**

When conducting quality improvement, I follow the instructions of my boss: "First, create an Ishikawa diagram with the others involved." However, despite the huge amount of man-hours required, I cannot find any factors that lead to a solution.

A6

SOUTH-FLOW assembles ke people early in the improvement. Then, after visualizing the facts through various multifaceted investigations with the strongest members, we will sort out the factors such as the Ishikawa diagram. When improving chronic defects, SOUTH-FLOW never engages in activities such as creating a Ishikawa diagram with only the initial members before conducting a factor investigation.

Q7 **I was told "we have made all possible improvements" and improvement does not progress. [★★]**

A7

SOUTH-FLOW assembles key people early in the improvement. Then, after visualizing the facts through various multifaceted investigations with the strongest members, we will sort out the factors such as the characteristic factor diagram and examine "possible factors" including the entanglement of factors.

Furthermore, we conduct experiments to find the optimal conditions for "contribution factors." There are many cases where the intertwining of these factors is the main point that could not be resolved until then.

Q8 I feel isolated from the in-house factory quality improvement support. [★★★]

I was ordered to dispatch to a factory to support quality improvement activities, which was to break through the current situation, but the factory staff at the dispatched destination did not welcome me.

A8

If you form the strongest team while paying attention to the points to keep in mind when using SOUTH-FLOW in Section 1.6 and proceed with the activities according to the SOUTH-FLOW scenario, the improvement activities will produce good results without making you feel isolated. For high-priority themes, we can obtain the cooperation of related parties by utilizing SOUTH-FLOW with less psychological friction.

Q9 The site is busy and we cannot get the cooperation of data collection. [★★★★]

I decided to use SQC to improve chronic defects with high priority, but I was refused because the production site was busy and I didn't have time to collect data for improvement.

A9

With SOUTH-FLOW, once the priority survey determines the chronic defects that should be addressed, the reliability of the quality evaluation is first confirmed. Specifically, it's just as if you are evaluating the quality of good and defective products on a daily basis. It's a simple but very important thing, and even on a busy site, they will cooperate, thinking, "I will cooperate to that extent." In SOUTH-FLOW, we maintain the centripetal force while gradually increasing the difficulty from the survey request that is easy to start.

Q10 The assignment of the orthogonal experiment was perfect, but the evaluation failed. [★]

I learned the design of experiments before and conducted an orthogonal experiment, but the trend of the factors was not seen as expected though it should be theoretically correct. Later, I learned that the evaluation method was unreliable, but it was too late.

A10

With SOUTH-FLOW, once the priority survey determines the chronic defects that should be addressed, the reliability of the quality evaluation is first confirmed. Specifically, we only ask them to evaluate good and defective products as if they were quality-evaluated, but this is very important and can never be done after an experiment.

Q11 Can a person work as an improvement leader without specific technology? [★★★★★]

In my second year at the company, I don't think I have acquired any unique skills yet, but I was appointed as a quality improvement leader. I'm in trouble because I don't know what to do as a leader.

A11

If you form the strongest team (including a key person with the skills of someone in the Union of Japanese Scientists and Engineers) while paying attention to the points to be noted in utilizing SOUTH-FLOW in Section 1.6 and proceed with the activities according to the steps of SOUTH-FLOW, you will be the person who promotes improvement. Even if there is a unique technology related to the defect, the entire improvement team can solve it.

Q12 I want to take over the improvement, but I am not good at filing. [★★]

A veteran manufacturing chief who is said to be an improvement professional will leave the company at retirement age. He wanted to take over the many improvements he had made so far, but the files aren't organized and it is probably not possible. In the future, I would like to pass on the improvement activities I carry out to the next generation, but I am also not good at filing and don't know what to do.

A12

The SOUTH-FLOW-equipped improvement software has an automatic filing function. A folder is automatically created for each improvement theme and the surveyed contents are automatically stored for each step of SOUTH-FLOW. Therefore, there is no need for filing.

Q13 In long-term improvement activities, it is easy to forget what we learned before. [★★]

Since it takes time to improve chronic defects, the content of the findings at the beginning of the improvement often fades from the evidence of the people involved, which is a problem.

A13

The SOUTH-FLOW-equipped improvement software has a Rolling paper and analysis result list function. The graphs and charts of the survey results can be stored on the Rolling paper so that you can see them at a glance for each step of SOUTH-FLOW. In addition, the "Action Panel" of each step also displays a large list of analysis results for that step. Therefore, even in activities to eliminate chronic defects, which tend to be relatively long-term activities, it is always possible to review the survey results.

Q14 QC circle resumes do not help improvement. [★★★]

I am entrusted with the manufacturing site as the group leader, but I think that the resume of the QC circle is for summarizing, and I do not feel the need for it during improvement.

A14

The SOUTH-FLOW-equipped improvement software has a keynote (plan and report) function. This keynote is created in an excel file. By pasting and editing important parts of the information stored on the Rolling paper into a keynote containing each step of SOUTH-FLOW and the procedure of factor investigation, you can report the progress even in the middle of improvement without hassle.

Q15 There is no end to the announcement of improvements to the QC circle. [★★★]

I am the secretariat of QC circle activities, but instead of proceeding with the circle systematically after finding a problem for themes that have already been improved and have a large effect amount (such as those that have hit ideas for improvement, not QC stories), the themes that were planned and advanced from the beginning when it was about to be announced, they are announced afterwards in the registration, which is a problem.

A15

The SOUTH-FLOW-equipped improvement software has a progress review list function. During the period of improvement, which survey items have been surveyed (a check mark is automatically added when the survey results are registered) or have not been surveyed in the procedures within the steps of SOUTH-FLOW. It is a list where you can see if it has been investigated or not in one glance. Therefore, the activity status of the improvement team can be easily viewed by others, and by utilizing this progress review list, it is possible to provide timely guidance on how to fill in the postscript during the activity period.

1.6 Points to Keep in Mind When Using SOUTH-FLOW

There are a few things to keep in mind when using SOUTH-FLOW. Here, we will introduce the following four points.

(1) The importance of confirming its effect within the company

(2) The importance of finding the cause of defects

(3) The importance of sampling

(4) The importance of the "2-6-2 Principle"

Many people take these points for granted, but they are all indispensable for improving chronic defects, so please reacquaint yourself with them.

(1) The importance of confirming the effect of SOUTH-FLOW within the company: preventing defective products from being released to customers

Improvement activities are centered on the concept of preventing the release of defective products, based on the quality improvement fundamentals of "don't introduce defective products into the next step in production" and "the next step is the customer." There are various improvement methods, including a "QA matrix," "QA network," and "quality chart" to reassess quality assurance systems and improve processes in order to prevent the release of defective products to customers.

Among these quality improvement methods, the SOUTH-FLOW process takes another approach of preventing the incidence of defects: "don't create defective products." Therefore, when conducting quality improvement activities under SOUTH-FLOW, the first step is to take measures to prevent the release of defective products to customers so that this does not occur during the improvement making period. We must make progress in our improvement activities without adversely affecting the customer.

Figure 1.6 is an example of an activity for preventing the release of defective products to customers. The calendar shows the number of days without customer complaints. When conducting activities to improve chronic defects, you should never use customer evaluations as confirmation of the effects of SOUTH-

FLOW.

(2) The importance of finding the cause of defects

Among those with experience in quality improvement, there are many who feel that "in quality improvement, once you have an adequate grasp of the current state of affairs, 80% of the work has been done." In practice, once the cause of defects is known, it does not seem to be so difficult for the engineers and other parties in charge to come up with countermeasures.

On the contrary, if there is only one cause of chronic defects and the technicians know what it is, then most companies will have already taken measures to resolve the problem.

In all of the quality improvement efforts the author has been involved in thus far, over 80% improvement has been reached if, upon investigations with key personnel to visualize (mieru-ka) the facts of the situation, the "contributing

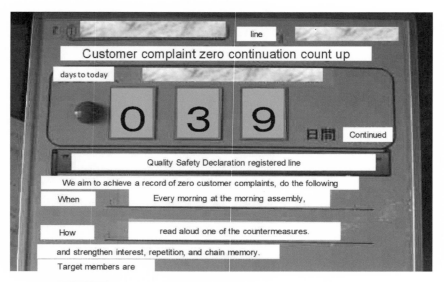

Figure 1.6 Example of an Activity for Preventing Defective Products From Being Released to Customers

factors" for defects could be determined from the "probable causes."

On the other hand, in the case where you have to experiment with factors pulled out from your own head, your "possible factors" will lack applicability to the situation, and in the end, despite having spent time and man-hours, the chronic defects will not go away. We have witnessed this.

There are various methods for SQC, but in the end, what is important in problem solving is what factors of the causes (probable causes) you experiment with, and whether or not these factors can be found by the people involved.

(3) The importance of sampling

If you take quality control training, you can learn many statistical analysis methods, from the QC Seven Tools (Q7) to experimental design. Sampling is one of the major differences between "analysis" as it occurs in training and "analysis" as it occurs in quality improvement in the field.

The data handled in training is nothing but a part of the data representing the population we want to compare. Since we test and estimate the difference in this population with the statistics, it is a prerequisite that all the data handled is obtained from a sample representative of the population.

On the other hand, for quality improvement in the field, it is necessary to confirm whether the obtained data is really representative of the population. Especially when it comes to improving chronic defects, it may be necessary to use sampling methods that have not been noticed before.

Therefore, when making improvements under SOUTH-FLOW, the sampling method is confirmed at an early stage of improvement to ensure correct sampling. Even in daily life, when tasting soup cooking in a pot, it is necessary to thoroughly mix the contents of the entire pot in order to get a representation of the soup's taste as a whole, rather than seasoning based on just tasting from the top.

The same principle applies to activities aimed at eliminating chronic defects; it is necessary to thoroughly take sample representatives of the population we want to compare from the field. We must proceed with making improvements while paying constant attention to the reliability of our sampling.

(4) Importance of the "2-6-2 Principle"

As mentioned previously, the "2-6-2 Principle" is indispensable for improving chronic defects in a company. Below, we will briefly introduce how to push forward improvement efforts upon understanding the "2-6-2 Principle".

If you were to carry out quality improvement activities in a team with ten members, those ten people could be roughly divided into three groups.

i) First group (20%): Ambitious, cooperative, positive types

There are two people in the first group, and the people who belong to this group are eager to work on everything, and will eagerly volunteer to cooperate in improvement activities.

ii) Second group (60%): Centrists depending on the atmosphere

The second group is a group of six people whose level of cooperation is hard to predict. Depending on the atmosphere around them, they may be cooperative or uncooperative. They are influenced by the members of the other two groups, including you, the promoter of improvement.

iii) Group 3 (20%): Contrarians? Uncooperative negative types

In contrast to the first group, for the two members of this third and final group, it takes a long time until they will cooperate, or they may even withhold their cooperation to the very end.

The previously-displayed Figure 1.4 shows the atmosphere of improvement based on the "2-6-2 Principle." As you move forward with your improvement activities, you should not hesitate to make an ally of the first group and thereby gain momentum. And as shown in Figure 1.4 (a), the shortcut to success for improvement activities is to make the utmost effort to create a workplace atmosphere that causes the second group to naturally go along with those around themselves and cooperate.

When you set about making improvements or making a new improvement theme in a new workplace, there are bound to be people you have never worked with before. Therefore, in order to make improvements everywhere, this "2-6-2

Principle" is one of the ideas that you should not forget as you move forward.

─── **Coffee Break** ───

(1) First Luncheon at a Workplace Under Improvement

Question: The first workplace you are put in charge of consists of the following staff: two men in their fifties, four in their thirties or forties, and four in their twenties, along with five women in their thirties and eight in their twenties, for a total of 23 people. You have been instructed to push forward their improvement activities. What actions do you take to promote the communication necessary for improvement in a short period of time?

▶ Example Answer

We plan a lunch for everyone in the meeting room as shown in the figure below. I's a good idea to get to your seat early and observe how the people get together, where they sit, and how they talk during the meal.

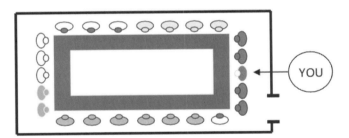

You can usually get a rough idea of the group structure of the work-place because people usually get together with friends and avoid eating with people they don't get along with. It is also a good idea to listen to people's conversations to find out who belongs to which group from the "2-6-2 Principle." Keep in mind that the shortcut for improvement is in creating an atmosphere.

Chapter 2 SOUTH-FLOW
How to Promote Chronic Defect Elimination Activities by SOUTH-FLOW
-Actual Quality Improvement by SOUTH-FLOW-

Chapter 2 introduces the procedures, procedure points, and improvement cases for each of the 11 SOUTH-FLOW steps. Each step has procedures and examples of improvements in eliminating chronic defects. Please check each step while paying attention to the points to keep in mind when using SOUTH-FLOW introduced in Chapter 1. In addition, all procedures have a "Main Points of the Procedure" column, which is a supplementary explanation to deepen the understanding of each procedure. Furthermore, in the improvement example, Mr. Minamiyama (the lead problem solver) introduced improvement activities that he was involved in at his first workplace and with the two improvement members of the factory to eliminate chronic defects in quality characteristics.

Chapter 2 introduces the contents of each step of SOUTH-FLOW in the configuration shown in Figure 2.1.

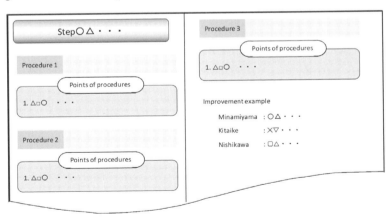

Figure 2.1 Configuration diagram of each step in Chapter 2

Step 1 Problem-Solving Solutions
[Scenario selections for pursuing the root cause]

Step 1 is divided into three procedures.

[Procedure 1] First, select the solution of the theme and then

[Procedure 2] Register the theme, clarify the reason for selection, and

[Procedure 3] Register further improvement members.

Each procedure of Step 1 is explained below.

[Procedure 1] Theme resolution scenario selection

(1) Selection of a scenario that resolves the gap between the original state and
the current situation is the most important thing in advancing improvement
to select an appropriate improvement scenario.

A problem is defined as "a gap between the ideal state and the current situation."

When the ideal state is "the state that it should be," we have to consider the
gap between the ideal and the current situation from the perspective of maintaining the quality of the manufacturing site. Naturally, we only want to make
good products, but at the present, defective products are being produced. This is
exactly the gap, and that is the problem at the site.

(2) In order to find (mieru-ka) the root of the cause of the defect hidden in the
site,select the quality improvement scenario to navigate improvements.

If a defective product occurs where a good product should be manufactured
on site, there is always a cause to the defect. In other words, the information
(the key) that causes it is always hidden in the field. SOUTH-FLOW is a quality
improvement scenario When defective products are mixed in the manufacturing
process where only good products should be produced, SOUTH-FLOW is a
quality improvement scenario on how to proceed with improvement activities
while bringing together the strongest members in the company. While statistically judging the information (data) from the site with the strongest members,
we will repeatedly ask why, why, and why to pursue the cause of occurrence and

connect it to the problem solving activity.

How to Proceed with Elimination of Chronic Defects by SOUTH-FLOW

Points of the Procedures

(1) Are there any defective products?
- Take advantage of SOUTH-FLOW when defective products occur.
- Are you trying to solve defects using QC methods?

[Procedure 2] Registration of Theme Name

If a defect occurs in [Procedure 1], select a centripetal chronic defect improvement theme in [Procedure 2].

Procedure Points

(1) Decide on a theme name.
- It should be concise so that you can understand the purpose of the content. For example, a theme name could be "To Eliminate Defective Dent of Product A" (at this point, you can tentatively decide).
(2) The reason for selecting the theme is also specified.
- Clarify the points (let's confirm this in Step 3).

[Procedure 3] Registration of Improvement Members

After clarifying the theme name and the reason for the theme in [Procedure 2], register the improvement members in [Procedure 3]. Elimination of chronic defects cannot be done alone. In particular, it is essential to make improvements with the understanding of your boss.

Procedure Points

(1) Register improvement members.

- Did the members who are enthusiastic about improvement come to-gether?

(2) Did you choose an improvement leader?

- Lead positive activities (leaders especially should always be positive).

(3) Do you have an adviser?

- Are your bosses and staff aware of improvement themes and members? (In Step 3 or 4, let's get comments from your superior)

[Improvement example 1] Product A Dent defect improvement -1
〜 "Problem Solver" Minamiyama Team's Record of Struggle 〜

This happened at sheet metal processing manufacturer A. At the manufacturing site, there were endless defects in new products that started mass production in January. With the end no where in sight, the defect rate is showing an increasing trend day by day, and if nothing is done, the amount of loss due to defects may eat up profits. It was only a week ago that Mr. Tokai, who had lost his temper, called Mr. Kitaike, the chief of the manufacturing section, and Mr. Minamiyama, who was in charge of improving the headquarters.

Tokai: Mr. Kitaike, can you explain what this means? As you probably know, our credibility with our customers has already fallen. If the number of defects continues to increase, the loss on defects will not only gobble up the profits, but it may also have a serious impact on the next order.

Kitaike: Well, I'm fully aware of that, but I haven't figured out the cause of the defect yet. We are also requesting support from the Quality Control Division, but in this situation where we are rushing to respond to complaints from customers, the investigation of the cause has not progressed at all. The site is kept busy with daily production, and it is all I can do to keep the water at bay. What should I do... I'm sorry.

Tokai: Yeah, it looks like you're having a lot of trouble. Of course, I don't think that management can just stand and watch this situation. Actually, I'm thinking of setting up a quality improvement project for this new product with Mr. Minamiyama, who was specially appointed by the general manager as the leader. But I wonder if he can get along well with everyone. It's the first time for Mr. Minamiyama to handle this product, so your cooperation is indispensable. Mr. Minamiyama, I'm counting on you.

Minamiyama: Well... yes, I'm not confident to be honest, but if I can be of help I'll do my best.

Tokai: Thank you, I will spare no effort to support you. I hope that the true cause will be clarified and improved by statistical analysis. It will be tough, but I'm looking forward to working with you. Anyway, as you can see from Figure 2.2, the number of defective Dent defect* is on the rise. Product A, a new product mass-produced in January, exceeds the defect rate upto 1.5%. As the first goal, I would like you to somehow reduce this to 1/10, that is, 0.15%. If the root cause is understood and the improvement is successful, we can apply the method to improve the dent defect to the other two varieties (horizontal deployment). The deadline for the first goal is the end of June, which is three months, and please do your best.

*Dent defect: A dent made by inserting a foreign substance into a processed sheet metal part.

[Birth of an improvement project team]

Under the above circumstances, the quality improvement project was launched. The members are:

Leader Minamiyama: This is his second year at the Kaizen (Improvement) Promotion Office of the Kaizen Activities Headquarters. He is a

mid-career employee who travels around each factory like a fire extin-
guisher at a manufacturing site where many defects occur, as a so to
speak "problem solver."

Chief Kitaike: A great veteran with 30 years of sheet metal processing experience. Although he is a stubborn craftsman, he is the most trusted
manufacturing person in the manufacturing professional field who is in
charge of this manufacturing site in general.

Nishikawa: He is the leadman at the manufacturing site, who is familiar
with everything from materials to equipment, and has an outstanding
QC sense.

As the above three people were specially appointed by the general manager,
they were allowed to start off with devoting almost 100% of their business hours
to this quality improvement project. However, after a week, the chief Kitaike was
finally starting to open up, and there was no sign of further progress. As expected,
the leader, Minamiyama, was also beginning to get impatient. The only line of
recourse was the improvement scenario "SOUTH-FLOW," which the headquarters' improvement promotion secretariat began using two years ago to eliminate

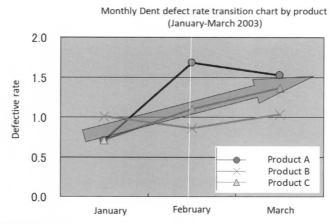

Figure 2.2 Monthly Dent Defect Rate Transition Line Graph by Product

chronic defects.

The three members of the improvement team immediately selected the quality improvement scenario because the yield was poor due to quality defects in [Step 1] according to [Procedure 1] of SOUTH-FLOW. In [Procedure 2], the theme name was set to eliminate defective Dents of product A, and the reason for selection was improvement by a special project from the general manager. In the member registration of [Procedure 3], they registered the three of themselves and started as advisors of the Tokai headquarters.

SOUTH-FLOW How to Promote Chronic Defect
Elimination Activities by SOUTH-FLOW

Step 2 Stratification for Priority Survey
[What to Start With]

In Step 1, you clarified the theme name and the reason for selecting the theme, and gathered improvement members. Step 2 conducts the first survey of focused activities to spend valuable time in the enterprise to eliminate chronic defects.

Step 2 is divided into three procedures.

[Procedure 1] By checking the occurrence status, select which theme should be solved

[Procedure 2] Set goals

[Procedure 3] In order to prevent defective products from leaking to customers during the activity period of improving chronic defects, check the warranty system is ensuring they are kept as internal defects.

Each procedure of Step 2 is explained below.

[Procedure 1] Confirmation of occurrence status

First, check the occurrence status. After clarifying which quality defect should be tackled, you will proceed with the investigation for improvement.

Main Points of the Procedure

(1) Check the occurrence status.
  - Investigate the frequency of occurrence and defect loss, and investigate which defects should be tackled with valuable improvement time to be efficient for the company.
(2) Is the problem you are having trouble with happening now?
  - Did the people concerned recognize the priority issues in the workplace?
(3) Useful QC Methods / Analysis Tools
  - Pareto charts, pie charts, etc.
(4) Particularly Useful QC methods
  - Figure 2.3 shows an example of an application that uses a Pareto chart to see what can be improved efficiently when conducting a priority survey.

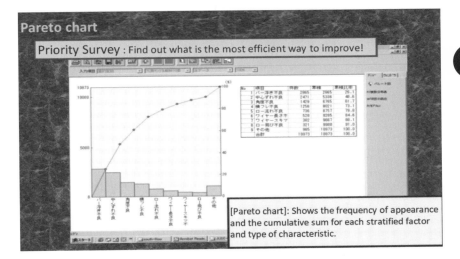

Pareto chart

Priority Survey : Find out what is the most efficient way to improve!

[Pareto chart]: Shows the frequency of appearance and the cumulative sum for each stratified factor and type of characteristic.

Figure 2.3 Pareto Chart for Priority Survey

Chapter 2

SOUTH-FLOW How to Promote Chronic Defect Elimination Activities by SOUTH-FLOW

[Procedure 2] Goal setting

After deciding the quality characteristics to be tackled in [Procedure 1], set the target value for eliminating chronic defects in [Procedure 2]. Basically, aim for "zero," and if it is difficult, aim for "a digit difference quality (1/10)." It is better to set an attractive value than "half the year" for the centripetal force of the activity.

Procedure Points

(1) Set target values for activities based on the results of the survey.
- When setting goals, clearly describe the management characteristics and target values by when they should be done (deadline).

[Procedure 3] Preventing the outflow of defective products to customers.

If the target value for prevention of occurrence can be set in [Procedure 2], in Procedure 3, when a chronic defect occurs, first of all, a device to prevent the chronic defective product from leaking to the customer (outflow prevention measure to stop it) is implemented.

If even one defective product leaks to the customer during the improvement period, not only will it cause trouble for the customer, but many people in the company will be forced to deal with it, and it will no longer be time for preventative measures. While stopping defective products in-house without causing any inconvenience to customers, it is far more psychologically viable to improve and carry out activities to eliminate chronic defects. Therefore, we should never have a system in which customers evaluate the effectiveness of the measures we will take.

Procedure Points

(1) Are the measures to prevent outflow to customers perfect?
- Let's establish a state where no defective products are sent to customers and concentrate on solving in-house defects.
- Do not make the customer an evaluator of improvement.
- Clarify what should be done in each process so that the defect will not be leaked to the customer.
- Check the quality assurance system for customers using the QA network, QA matrix, quality charts, etc.

(2) Useful QC Methods / Analysis Tools
- Matrix method (QA network, quality chart, etc.)

[Improvement example 2] Product A Dent defect improvement-2 [Difficulty of in-house improvement]

Kitaike: Hey leader, what are we going to do now...? For the time being, I chased the defective data for a week, but all I found was that there were still many defective Dents. But I knew that a long time ago.

Nishikawa: Kitaike, it's needless to say, it will be a total loss. We are gathering to find out the cause, so the problem is what to get from the on-site data. Kitaike: Don't be silly. That goes without saying. So what does Nishikawa know from this data?

Also, I think that the Ishikawa diagram that you learned how to do during this time was made with the help of everyone in the field, and I think you took a lot of man-hours to make it.

Minamiyama: Both of you stop breaking this group up. Yes, as Kitaike said, we took a week to rearrange the transition chart of the number of defects over the past three months as shown in Figure 2.2 (above) and to investigate the types of defects as shown in Figure 2.4. It was found that defective Dent accounted for 73% of all defects, 66% of them occurred in the R part of the product surface, and the defective rate of Dent in Product A was the highest. It's my first product and manufacturing process, so I'm sorry to have taken your time to teach me the manufacturing site.

Kitaike: Well, that's right, leader, even if you're a problem solver, it's not easy for the first product. Most of the time, bad Dent is caused by heavy burrs and debris getting into the mold. The only thing I'm worried about is that the goal of improving the incidence of dents to 1/10 (up to 0.15% in 3 months) is too high.

Pareto chart of product A by defective item (January-March 2003)

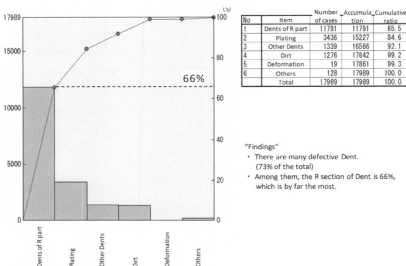

No	Item	Number of cases	Accumulation	Cumulative ratio
1	Dents of R part	11791	11791	65. 5
2	Plating	3436	15227	84. 6
3	Other Dents	1339	16566	92. 1
4	Dirt	1276	17842	99. 2
5	Deformation	19	17861	99. 3
6	Others	128	17989	100. 0
	Total	17989	17989	100. 0

"Findings"
- There are many defective Dent.
  (73% of the total)
- Among them, the R section of Dent is 66%,
  which is by far the most.

Figure 2.4 **Comparison of Product A by Defective Item Pareto Chart**

[Conventional Problem-Solving Difficulties]

Nishikawa: If you can't identify the cause of the burrs and debris, you can't do anything about it. Why don't you review the Ishikawa diagram we made the other day about "Why are there so many Dent defects?" for factor analysis.

Kitaike: You don't know anything! Do you think I would do something which took so much time and did not lead to any improvement again? Thinking about it now, some circle should have already solved it if you put in that kind of work.

Don't you know that? After that, there was a girl, I think her name was Nakamura, she went home crying with bright red eyes. She said, "I'm being doubted as a worker, even though it's about the improvement of Dent." The next day I apologized to her on your behalf!

Nishikawa: Don't blame me so much. I'm using all my knowledge to think as hard as I can...but I don't think it's useless to speculate about the cause with 4M for the time being, what do you think?

Minamiyama: Yeah, you made it so you might as well show it to me.

Kitaike: Hey, leader, are you serious? I'm so busy at the site that I'm tired of seeing things that take such a long time. I'll go back to work if you ask me to do that again.

Minamiyama: Wait a minute, let's do this. According to the SOUTH-FLOW recommended by the headquarters for chronic defect elimination activities, the Ishikawa diagram is for after grasping the current situation. So forget about the Ishikawa diagram for the time being and organize the current situation according to SOUTH-FLOW recommended by headquarters. First of all, in "Step 1" the question is "Is there a quality defect?". So in the case of our project this time, the answer is YES. Next, regarding "confirmation of occurrence status" in "Step 2," the number of defective Dent is the highest at 73%, so you should concentrate on this.

Kitaike: That's right, and I also found that 66% were concentrated in the R department. You're asking a simple question, isn't this Southern thing somehow fooling us? The problem is that it's not going any further at all.

Nishikawa: Minamiyama, the Dent in the R section is like the picture in Figure 2.5. At the customer's site, the other part is in close contact with this part, so it seems to be a problem if there is a Dent.

Minamiyama: It's a precision press product, isn't it?

Nishikawa: Isn't it better to focus on investigating the process of performing this R molding?

Minamiyama: That may be true. However, SOUTH-FLOW is systematized to investigate the site from various angles and the relationship between the processes before and after, so let's investigate according to the scenario.

Kitaike: Nishikawa, you know, the R molding process hasn't been reduced even if the mold specialists have investigated it so far, and there's no shortcut to improve it.

Nishikawa: You don't have good ideas either, Kitaike.

Minamiyama: Well, this is not the time for international conflicts, both of
you.

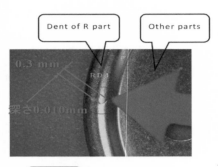

Figure 2.5 Dent Defective Photo

─────────── **Coffee Break** ───────────

(2) High-pressure street sales and quality improvement

▶ Question:

Which question is easier to answer if someone tries to sell you something on the street?

A. 「Excuse me, what time is it now?」

B. 「Excuse me, do you want to go abroad to learn English??」

A. "Excuse me, what time is it now?"

B. "Excuse me, do you want to go abroad to learn English?"

▶ Example answer:

Most people's answer will be A. Questions A and B are both the first questions of high-pressure street salespeople at English schools that recommend studying abroad in the United States. So why do so many people choose A? There are two main reasons:

(1) Question A is a short, simple word that is easy to understand (less resistance). On the other hand, Question B is the opposite and difficult to understand.

(2) Question A is easy to answer (it is easy to get interested and a kind response). On the other hand, Question B is difficult to answer at once.

These points are the same for quality improvement and quality education, and are important for attracting stakeholders in the early stages. Activities that are simple, easy to understand, and lead to a kind of charm are important.

For example, when requesting a factor investigation from the person concerned, rather than requesting cooperation that takes time to create an Ishikawa diagram, ask "Would you like to measure defective and non-defective products several times?" or "How do you judge this quality characteristic?". It is easier for the other party to answer.

Step 3 Quantification of Defect Characteristics
[Reliability confirmation of quality characteristic evaluation]

Once the high priority and chronic defects to be addressed are selected in Step 2, the next step is to check the reliability of how to evaluate the quality characteristics.

Step 3 is divided into three procedures.

[Procedure 1] Measure the quality characteristics of non-defective and defective products and confirm that the evaluation is reliable.

[Procedure 2] If in doubt, check how many times the same measurement is repeated and the average value is considered to be the true value.

[Procedure 3] Especially when using a measuring jig whose measurement method is determined in the drawing, use the gauge R & R (GR & R) method to check the reliability of the measurement against the quality characteristic standard.

These studies confirm the measurement, which is one of the 5M's, in Step 3, which is the earliest stage of improvement.

If the quality characteristics to be improved use a measuring jig, it is recommended to skip [Procedure 1 and 2] and perform [Procedure 3].

Each step of procedure is explained below.

| [Procedure 1] | Measure the quality characteristics of non-defective and defective products |

The simplest and most important check is to evaluate the good and bad products several times. At this point, if you can confirm the phenomenon that a good product was evaluated as a defective product at one time and vice versa, it means that the data used for the priority survey in Step 2 cannot be used. In that case, you will need to start over from Step 2.

Procedure Points

(1) Repeatedly measure good and defective products.

- Is the judgment the same even if good and defective products are repeatedly measured? (Remove the object from the measuring machine after each measurement)

(2) Are you confident in the value (judgment) no matter how many times you measure it?

- If the target of the defect is a sensory test such as a visual inspection, check the reliability of the judgment, such as using a limit sample or accrediting an inspector.

(3) The measurement method may not be catching up with the improvement of customer needs

- Are you clearly judging "defective/non-defective" and "defective and good"?

(4) Measure the same product 5 times or 10 times to check the reproducibility of the measurement.

- Let's compare the standard width and the repeated measurement confirmation result.

(5) In the case of destructive inspection, pay particular attention to the reliability of the judgment.

- Measured values and reproducibility may differ depending on the measurement method. Future survey data must be reliable.

(6) Is it necessary to have the key person of measurement participate?

- If there is a problem with the measurement, involve the person in charge of the measurement.

(7) Useful QC Methods / Analysis Tools / Line Graphs, Bar Graphs, etc.

[Procedure 2] **Perform multiple measurements (larger sample sizes) and check the true value**

If the reliability of the measurement is low in [Procedure 1], you can use the future analysis data by considering the average value of the data obtained by repeatedly measuring the same thing as the true value. For example, when measuring quality characteristics such as coating film thickness of a product with a film thickness meter, it is easier to find the true value when measuring the same product several times than when measuring the data once.

(Procedure Points)

(1) How many times can you trust the average value of the measured values?
- Use the law of large numbers (the average value measured multiple times (larger sample sizes) approaches the true value) (see Tea Time (3)).
(2) Useful QC Methods / Analysis Tools, Line Graphs, etc.

[Procedure 3] Implementation of gauge R & R (GR & R)

GR & R here is a method of statistically evaluating the reliability of whether or not the current measurement method can correctly evaluate the product's performance with respect to the product standard. Even with a measuring machine that has been calibrated on a regular basis, quality evaluation problems may occur depending on the quality characteristics of the product you want to measure and its standards. Furthermore, depending on the product to be measured, quality evaluation problems may occur due to the quality of the fixing jig and the difference in the measurer. For details on GR & R, refer to Chapter 3, Section 3.1.

(Procedure Points)

(1) Have you decided on the measurement method and measurement jig?
- If the dimensional measurement method and its measuring jig have been decided in advance, try performing a gauge R & R (GR & R) that can evaluate reliability.
(2) Register on the Rolling paper and organize the main points in the keynote.
- Let's keep a record of the progress of activities little by little.
- At the same time, let's recall the results of the survey so far.

- This is very convenient when putting everything together later. In particular, "interest," "repetition," and "chain memory" are indispensable for discovering factors and countermeasures.

(3) There is measurement accuracy of quality characteristic values.

- If you have multiple measuring machines, check the repeatability, people, and machine differences.
- Furthermore, measure several samples at the upper, middle, and lower limits of the standard, and perform a correlation survey.

(4) Improvement of robustness by processing measurement data

- If there are abnormal values in the measured data, consider measures to remove the abnormal values such as median ($\bar{x}$: median measured value).

(5) Useful QC Methods / Analysis Tools

- Gauge R & R (GR & R), correlation diagram, etc.

(6) Particularly Useful QC methods

- Let's utilize GR & R based on the concept of quantification of measurement as shown in Figure 2.6.
- Make sure that the measured value is reliable by GR & R.
- If you are not confident in the measurement content, how many times can you trust the average?
- Figure 2.7 shows an application example for confirming the reliability of measurement using GR & R.

GAUGE REPEATABILITY AND REPRODUCIBILITY DATA SHEET (Long Method)

[Improvement example 3] Product A Dents defect improvement-3
[Measurement reliability]

Minamiyama: Okay, let's check "Step 3" in accordance with SOUTH-

FLOW.

Nishikawa: Well...it seems that "Step 3" is "the quantification of defect characteristics," so it is necessary to confirm the reliability of the measurement.

In other words, it is important to make a judgment by a method that can give the same judgment result no matter how many times anyone makes it so that the quality judgment is not overturned by measurement variations and errors.

Minamiyama: That's right, you are quick to understand, Nishikawa. Certainly, there is no problem with this Dent defect because the level of the people concerned is adjusted with the limit sample and the certified inspector system is used, so the judgment of good or bad is maintained in a state of being almost 100% reliable.

Nishikawa: Yes. Now let's check Step 4 in accordance with SOUTH-FLOW.

Kitaike: Hey leader, you can check "Step 4," but do you know what process order this product is manufactured in?

Minamiyama: Of course, because Kitaike gave me the flow of Figure 2.8 when he first showed me around the site.

Nishikawa: Well, I guess Kitaike, you're not so old.

Kitaike: No, I was just testing the leader you know.

-Is the Characteristic Value Reliable? -

- Is there a difference when measuring the same product **"three times (repeatability)"**?
- Are there any measurement differences between **"measurers / measuring machines"**?
- Statistically judge the reliability of measurement → **"GR & R"**
  - (1) Measurement comparison of multiple samples (10, 20, 30 etc.)
  - (2) Measurement comparison with multiple measurers (by three people)
  - (3) Measurement comparison at multiple measurement opportunities (for three days)
  - (4) Comparison of reliability with respect to standard tolerance width
  - * Judgment: Good enough [≦ 10%], Conditionally good [≦ 30%]
- Data is uncertain: Utilization of **"average value of multiple measurements"**

Figure 2.6 Key Points for Quantifying Defect Characteristics

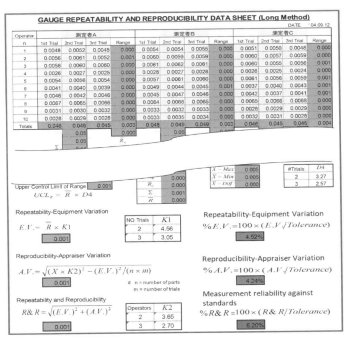

Figure 2.7 Gauge R & R (GR & R) sheet

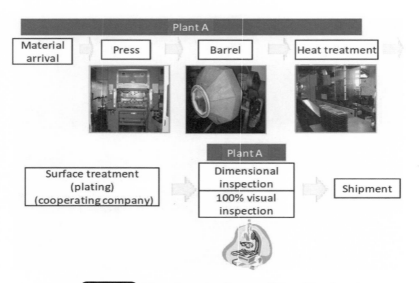

Figure 2.8 Manufacturing Process Flow of Product A

Step 4 Stratification for Factor Investigation (I) [Phenomenon]
[Finding factors from the result system]

After confirming that there is no problem with the reliability of the measurement in Step 3, in Step 4, we investigate the phenomenon side where the quality defect occurs, that is, the result system. In Step 4, we first search for the key people necessary for improvement, and then proceed to search for the cause of quality defects while forming the strongest team of the key people gathered.

Step 4 consists of [Procedure 1] to [Procedure 6]. Step 4 can be roughly divided into the first half and the second half as follows.

[Procedure 1 to 3] "First half": "Key person excavation and sampling method survey"

[Procedure 4 to 6] "Second half": "lot control and physical observation survey"

The "first half" of Step 4 is divided into three procedures.

[Procedure 1] Investigate trends by product and search for key people with similar products.

[Procedure 2] Investigate the fluctuation between manufacturing processes and search for the key people in the previous and next processes.

[Procedure 3] Then, investigate the time-series fluctuations within the same lot (same manufacturing conditions) and determine the sampling method so that statistical judgments in future activities will not be mistaken.

Furthermore, in the "second half" of Step 4, confirm the phenomenon in three procedures with the key people found in the first half.

[Procedure 4] Investigate trends between lots (different manufacturing conditions),

[Procedure 5] Investigate whether there is a trend in the defect occurrence site, and

[Procedure 6] Observe the actual product from various angles.

In Step 4, it is not necessary to carry out all the procedures according to the

Chapter 2 SOUTH-FLOW How to Promote Chronic Defect Elimination Activities by SOUTH-FLOW
-Actual Quality Improvement by SOUTH-FLOW-
Chapter 2 SOUTH-FLOW How to Promote Chronic Defect Elimination Activities by SOUTH-FLOW
-Actual Quality Improvement by SOUTH-FLOW-

theme of chronic defects to be eliminated, and we will proceed in consultation with the key person. If you cannot obtain the possible factors from the survey results so far in the factor sorting stage of Step 6, you may return to the skipped survey items later.

Each procedure of Step 4 is explained below.

[Procedure 1]   Is there a trend among each product?

Check if the defect occurs in similar products in the same way.

If it does not occur in a similar product, you should ask the person who is familiar with the product.

It means you could get cooperation in obtaining opinions on the information obtained in future surveys.

In addition to the original members, you have found the key people needed for improvement.

On the other hand, if similar products occur in the same way, it can be seen that horizontal deployment can be expected after this improvement.

Main Points of the Procedure

(1) Is it a defect limited to the product?
- If the defect is peculiar to the product shape, check if there is any difference in the shape, usage status, or manufacturing process from other products.
- Also, if it occurs in other products in the same way, it seems that horizontal deployment can be expected for other products with this improvement.

(2) The key is to discover differences by stratification
- Is it clear what to investigate, such as product shape and the manufacturing process?

(3) Is it necessary to have a key person of similar products to participate?
- If the trend differs depending on the product, ask the person who is fa-

miliar with the product to cooperate in asking why it seems to be different.

(4) The sample is a representative of "soup"

- Is the sample representative of the group you want to compare?

    Example) When you cook at home and taste the soup in the pot, you never evaluate only the top layer, but after stirring it, you taste a representative of the whole thing.

(5) Useful QC Methods / Analysis Tools

- Pareto charts, histograms, bar graphs, etc.

(6) Particularly Useful QC Methods

- Figure 2.9 shows an application example for checking the difference in the occurrence status of each product using a Pareto chart.

[Procedure 2] **Is there any variation between manufacturing processes?**

While [Procedure 1] confirms the trend for each product, [Procedure 2] confirms whether it occurs in the defective process or whether multiple processes are involved.

If a process other than the process in the field of the initial improvement member is involved in the defect, we will also ask people who are familiar with other processes. We have to request their cooperation in obtaining opinions on the information obtained in future surveys. If the investigation of [Procedure 1] and [Procedure 2] of Step 4 finds a key player who needs improvement other than the original members, it means the beginning of cross-functional improvement activities.

Main Points of the Procedure

(1) Where is the problem occurring in the entire process?

- Let's investigate in which process in the manufacturing process the quality defect to be improved occurs, and whether it is related to the pre- and post-processes.

- Don't be overconfident that your process is okay.

(2) The key is to discover differences by stratification

- Is it clear what to investigate from upstream to downstream of the manufacturing process?

(3) Is there a key person in the team for the process to be improved?

- If the original members do not have a professional in the process to be improved, let the other colleagues know the need for improvement and be sure to have them join. Even if it is difficult to ask directly, ask an adviser (boss or other staff) to help you request cooperation.

(4) Did you organize the strongest team in the company regarding the problem?

- This is the perfect setting for cross-functional improvement that transcends the boundaries of teams and departments.

(5) The sample is a representative of soup

- It is necessary to devise an analysis method depending on the sampling method.

- For F-correspondence, a multivariate relational diagram that investigates the same thing one-to-one for each process, and a comparison graph of statistical values such as mean value and variation (standard deviation) for sampling.

(6) Useful QC Methods / Analysis Tools

- Multivariate association diagrams, line graphs, bar graphs, etc.

(7) Particularly Useful QC methods

- Check the degree of influence of the manufacturing process that affects quality by using the multivariate association diagram as shown in Figure 2.10. Each is suitable for paired data processing.

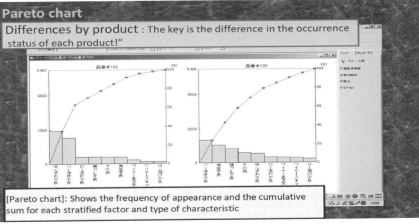

Figure 2.9 Pareto Chart for Checking Differentces by Product

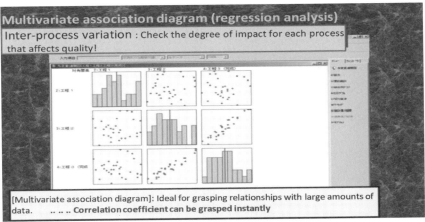

Figure 2.10 Multivariate Association Diagram for Inter-Process variation survey

[Procedure 3] Is there a trend for intra-lot fluctuations?

You have already searched for the key people in [Procedure 1] and [Procedure 2]. In [Procedure 3], you will investigate sampling, which is the basis of statistical analysis. Here, you have to check whether there is a trend in chronological order within the lot where the manufacturing conditions of the defect are not changed. If there is no noticeable trend, future research will be done with random sampling.

If trends are discovered, you can be aware that there are variables that you have not previously managed.

Since it is SQC to statistically judge whether or not there is a difference in the compared sample data, whether or not the sample is the group representative to be compared can be said to be the life of the reliability of the field data.

Main Points of the Procedure

(1) If the products are produced under the same manufacturing conditions, are the defect occurrence situations the same?
- Here, let's investigate the "variation" of intra-lot, paying particular attention to the time-series changes within the lot.
- Are there any trends in the product quality within the lot, such as from the start of production to the middle of production and the end of production under the same conditions?

(2) Useful QC Methods/Analysis tools, Line Graphs, Histograms, etc.

(3) Particularly Useful QC methods
- Figure 2.11 shows an application example using a line graph, which is effective for confirming trends even if the manufacturing conditions are constant.

Chapter 2

SOUTH-FLOW How to Promote Chronic Defect
Elimination Activities by SOUTH-FLOW

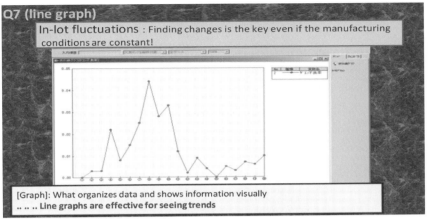

Q7 (line graph)

In-lot fluctuations : Finding changes is the key even if the manufacturing conditions are constant!

[Graph]: What organizes data and shows information visually
.. .. .. Line graphs are effective for seeing trends

Figure 2.11 Line Graph for Intra-Lot Fluctuation Survey

—— Tea Time ——

(2) Familiar "Stratification"

Even if we say "stratification," you might not get it. However, we have used this "stratification" in various situations since we were young and did not know the statistics.

We will introduce an example of how we have naturally acquired the idea of "stratification" from a young age.

-Insect Collecting "Beetles"-

First of all, remember back to when you went into the forest during summer vacation to collect beetles for your insect collecting assignment. If you have never experiences this, then try imagining it.

When we were young, we went to collect insects after being taught by books and parents and siblings that beetles gather in oak trees that suck sap at night. We also found them fallen under streetlights in the early morning using the knowledge that beetles gather where light is.

We used "classification / difference," so-called "stratification," without being aware of it. You learned that beetles are active "night in order to eat," so you chose "oak trees," a favorite of beetles in the forest and searched for "trees with dripping sap." Also, even if you did not enter the forest, you knew that the "collection points" were those such as "under streetlights in the early morning" instead of the paddy fields (except when there is a watermelon field or compost storage area nearby). This so-called "difference" is "stratification." You were collecting beetles by using it.

As a more efficient collection method, we also tied cotton soaked with a solution of brown sugar to trees to lure beetles who did not know where they were in a forest. When it was difficult to find more, this act of systematically setting a liquid similar to sap to catch beetles is exactly the experimental design idea in statistics.

-Pest Control "Cockroaches-"

If you think about it, even for the extermination of pests such as cockroaches that live indoors, you can set up "boric acid dumplings" or "cockroach killer" in the place where many cockroaches are found or in the shadow of their paths. It is also used to eliminate the fact that cockroaches hide in the shade around water and are omnivorous, preferring dried foods such as dried sardines. Even if you get rid of the cockroaches in front of your eyes, almost no one thinks they can get rid of the cockroaches inside the house. This is because you know from the habits of cockroaches that it is natural to think that if one or two cockroaches are found, there are about 30 lurking somewhere.

-Quality Defect Improvement and "Individuals"-

Now, when quality defects occur in the manufacturing industry, we try to find the cause. Encyclopedias and insect pictorial books have hints on when and how defects occur, what characteristics they have, and insect collecting and pest control. But when trying to improve quality defects, we have no choice but to rely on individual skills. "SOUTH-FLOW" is a quality improvement scenario that realizes "stratified" guidance for quality improvement for those who worry and think, "If only there were hints like the ones in cockroach extermination in activities to eliminate chronic defects" or "I wish quality improvement could proceed according to the QF story..."

[Procedure 4] Is there a trend for lot-to-lot fluctuations?

After confirming the key people and sampling method required for improvement in [Procedure 1] to [Procedure 3], which is the first half of Step 4, the lots (differences) whose manufacturing conditions have been changed in [Procedure 4] in the latter half of Step 4, check whether there is a trend among the status of defects and statistical values of quality characteristics (mean values, standard deviations, etc.). If there is no noticeable trend, it is possible that there are factors other than the conventional management factors.

In addition, when a trend is discovered, you can notice that there is a change in the management point that caused the fluctuation.

Procedure Points

(1) Does the occurrence of defects change when the manufacturing conditions are different (lot differences)?

- Lot-to-lot surveys, namely compare lots with different manufacturing conditions and investigate the effect of quality due to changes in manufacturing conditions.
- It is important whether or not the intra-lot fluctuation survey has already been conducted. Depending on the sampling method, there are many cases in which changes due to different lots of materials and parts are overlooked, or conversely, a misdiagnosis is made as if there were changes.

(2) Useful QC Methods / Analysis Tools

- Control charts, line graphs, histograms, etc.

(3) Particularly Useful QC methods

- Figure 2.12 is an application example that visualizes change points and trend changes when manufacturing conditions are changed using a control chart (for analysis).

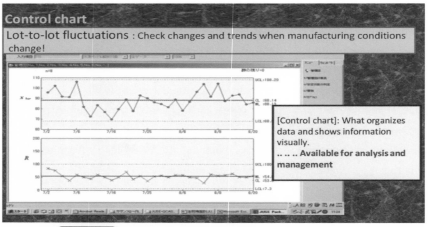

Control chart

Lot-to-lot fluctuations : Check changes and trends when manufacturing conditions change!

[Control chart]: What organizes data and shows information visually.
.. .. .. Available for analysis and management

Figure 2.12 **Control Chart for Lot-to-Lot Fluctuation Survey**

[Procedure 5] **Is there a trend in which part the problem occurs?**

Since we confirmed the on-site information about the change points be-tween lots in [Procedure 4], we will focus on the part where the defect occurs in this [Procedure 5]. If there is a trend in which part the problem occurs, the cause can be found relatively easily. Therefore, when improving chronic defects, it is extremely rare that this investigation alone will solve the problem. Because in many manufacturing industries, quality improvement is done daily, it is unlikely that any improvement will deviate from this survey point. In other words, it must have been improved and resolved before it became a chronic defect. However, when searching for the root cause of chronic defects, it is effective to sort out the factors in combination with other information.

Procedure Points

(1) If you find a trend in the defective parts, you grasp which points at the site you need to see in detail.

• This is the most orthodox investigation when investigating the phenom-enon. If you can find the trend of the site of occurrence by observing defective products, it is not so difficult to find the cause.

- There are almost no cases of chronic defects that can be solved by this single survey item, and there are many cases that can be solved by combining other survey items.

(2) Useful QC Methods / Analysis Tools

- Check sheets, photos, videos, etc.

(3) Particularly Useful QC methods

- Figure 2.13 is an application example using a check sheet to infer the cause from the characteristics of the occurrence situation.

[Procedure 6] **Is there a difference between non-defective products and defective products?**

Since the difference in the occurrence site was confirmed in [Procedure 5], in [Procedure 6], the defective product and the non-defective product are compared and observed except for their quality characteristics. For example, even if the quality characteristic of the defective product is the operating force, both the defective product and the non-defective product are cut and the cross sections are compared and observed. Figure 2.14 (a) shows the cross section of a non-defec-

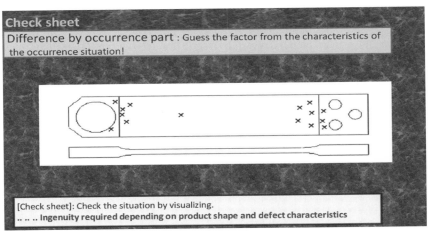

Check sheet

Difference by occurrence part : Guess the factor from the characteristics of the occurrence situation!

[Check sheet]: Check the situation by visualizing.
.. .. .. Ingenuity required depending on product shape and defect characteristics

Figure 2.13 Check Sheet for Investigating Differences by Part of Occurrence

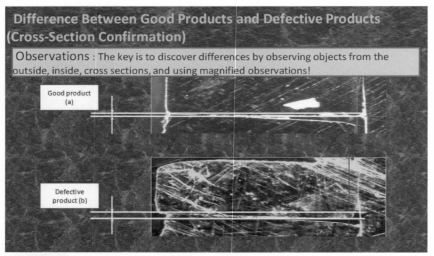

Figure 2.14 Cross-Section Observation for Finding Differences Between Non-Defective and Defective Products

tive product, and Figure 2.14 (b) shows the cross section of a defective product. The actual product does not have the dimension lines shown in Figure 2.14, but you may notice the difference in the size of the gap between the parts by looking at the cross section with the key people who were called in the first half of Step 4.

The strongest members convened in the first half of Step 4 will seek new discoveries, such as trying the methods that have been confirmed so far.

Procedure Points

(1) What is the difference between a good product and a defective product or a highly defective product and a slightly defective product?

- Let's find the difference by comparing them from various angles other than the quality defect characteristic.

    Example) Thoroughly compare defective products using possible methods such as external, internal, cross-section, and magnified observation.

- Furthermore, if there is a negative defective product and a better defective product in the quality defect characteristics, compare the three types including the non-defective products of the standard center.

(2) Goal Setting/Review of Activity Plan

- Is it necessary to review the target value and activity deadline from the survey results?
- We have all the people involved. Organize the division of roles and create a plan after a detailed survey.
- Reconfirm goal setting and activity plans with your boss.

(3) Useful QC Methods / Analysis Tools

- Line graphs, scatter plots, photographs, videos, etc.

[Improvement example 4] Product A Dent defect improvement-4
[New problem-solving scenario]

Minamiyama: You're right, as a result of properly adjusting the levels of the parties involved in this Dent defect, we have maintained a state of being almost 100% reliable in judging whether it is good or bad, so we cleared "Step 3."

Nishikawa: It seems so. Now we can go to "Step 4" "Strategies of Factor Investigation (I)." Here, it seems that defects are stratified from various angles in order to grasp where the cause is lurking, but this is one of the major features of "SOUTH-FLOW."

[Comparison With Similar Products]

Minamiyama: That's right, the first question is "(1) Is there a trend among each product?"

If there is a difference in the occurrence status of similar products, it means that the shape of the product may be related to the occurrence of defects. This is just a Pareto chart of the bad data we investigated last week.

Nishikawa: I know how to make a Pareto chart. It looks like this in Figure 2.15.

Minamiyama: As expected, Nishikawa, you're used to using the QC Seven Tools, and it's obvious when you line up the Pareto charts of three similar models.

Nishikawa: Well, it seems that there are the most Dents in the R department.

Minamiyama: What I can say here is that 70-90% of all three models have defective Dent, and all of them are concentrated in the R part. I wonder if it is still occurring in the mold?

Nishikawa: I can't say for sure, but that's a big possibility. Moreover, if we can improve Product A this time, it seems likely that we will be able to expand horizontally to the other two varieties. Is it only because of the feeling that we are one step closer to something different than before from this alone...?

Minamiyama: Now, the next item to check is...

Pareto chart by product / defective item

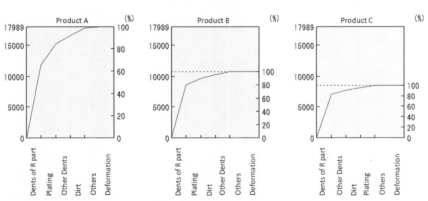

☆ In all three models, defective Dent accounts for 70 to 90% of the total, and all are concentrated in the R section.

☆ There is a trend in the part where Dent occurs → Does it occur in the mold?

Figure 2.15 Comparison of Occurrence Status by Product (Pareto Chart)

[Search for Processes to be Improved]

Nishikawa: Well, next is "(2) Is there any variation between manufacturing processes?" in Figure 2.16.

Minamiyama: Oh yeah, did you investigate the defective data for each manufacturing process?

Nishikawa: Yes, then there is data for the sampling survey of the defective rate of 100,000 pieces / lot we asked each process to conduct. What kind of QC method should we use?

Minamiyama: There are multivariate association charts and line graphs, but since we are looking at the transition of the defective rate step by step here, the line graph seems to be appropriate as shown in Figure 2.17.

Nishikawa: Wow, obviously there seems to be fluctuations in the defect rate between manufacturing processes. I didn't know it was like this.

Minamiyama: Especially when you see that the defect rate has decreased significantly in the barrel process, it seems that the problem will be narrowed down to the press process.

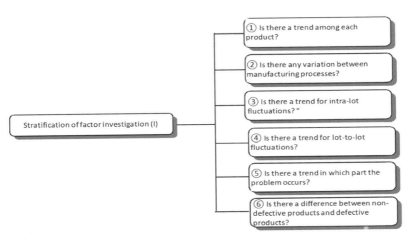

Figure 2.16 SOUTH-FLOW Factor Investigation Stratification (1) setp (Sysstem Diagram)

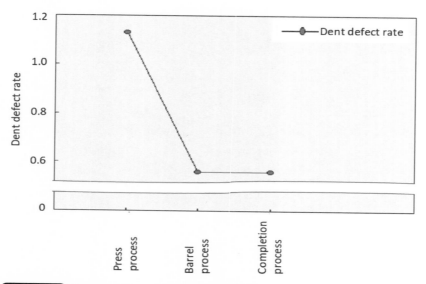

Figure 2.17 Line Graph for Checking the Occurrence Status Between Manufacturing processes

[Same Manufacturing Condition Survey (In-Lot Survey)]

Nishikawa: Well, in "(3) Is there a trend for intra-lot fluctuations?", let's narrow down the survey to the press process. Certainly, we have not conducted a time-series survey on lots with the same manufacturing conditions.

Minamiyama: It may take some time, but let's go after tomorrow's lot this morning. Oops, Kitaike just came back. Kitaike, how was the site today?

Kitaike: Hmm, defective products have begun to increase again...the people at the site are also desperate for sorting inspections to prevent defective products from leaking out. So what is the result of the "Computer Geek Team"?

Nishikawa: The "Geek" is unnecessary, Kitaike. However, it seems that I feel that I am one step closer to the root of the defect.

Kitaike: Wow, you are very confident.

Minamiyama: Therefore, I would like to ask Kitaike to help me carry out an important survey.

Kitaike: What on earth are you doing?

Minamiyama: I would like to follow the time-series survey of the defective rate of the press process for each product box.

Kitaike: How the hell did you come up with such a difficult investigation. But it sounds interesting, I've never heard of it before. It may take some time, but I may be able to get some clues. Anyway, there is no way to do it as it is. Do you follow the policy of the "Computer Geek" team?

Nishikawa: Like I said the "Geek" is unnecessary.

Kitaike: Stop your complaining, it's going to be a hard day tomorrow. It might take over a day and we won't finish.

[Accumulation of Small Discoveries and Formation of Teamwork]

By proceeding with the analysis according to the "SOUTH-FLOW" scenario, the Minamiyama team gradually began to demonstrate teamwork while gaining the feeling that the cause of the defect was approaching. The next day, it took a lot of effort to investigate the time-series changes in defects for 100,000 products per lot. With the cooperation of certified inspectors, they checked the defect rate of products that were sorted into 20 boxes for every 5,000 pieces. It seems that the members who finally returned to the office after completing the investigation when they entered the overtime hours got a response from each. In particular, Kitaike, who focused on the data around noon, was a little excited.

Kitaike: Hey Nishikawa, please graph this data right away, and I'm not sure, but obviously the appearance of defects was different between the morning and the afternoon. If you make a graph, we may see some trends. Now is the time to show off your "computer geek" skills.

Nishikawa: I'm not a geek, you know. Well, it seems that the products in the latter half of today that I investigated had relatively few defects, and as Kitaike said, if I graph the transition of defects, we may be able to understand something.

Minamiyama: That's right, we may finally be able to catch the root of the defect. Let's do it right away.

Nishikawa: Yes, I understand. Well, this way...

Kitaike: Don't rattled, just make a graph, silly.

Nishikawa: I understand, so you don't have to call me silly. Yes, it's done. Figure 2.18 is a time-series transition graph of the defective rate.

Kitaike: Oh, isn't it amazing Nishikawa? As expected, you are a computer geek. This is the increase in abnormal defects from (5) to (10). This can't be normal, there must be some change.

Nishikawa: That's right. I didn't notice it until now, and I didn't know that such strange changes have occurred in chronological order...on the other hand, after the afternoon (11) I investigated, the defect rate has settled down considerably, what has changed? What should I check first? The equipment, materials, people, and methods, is it 4M?

[Subtle Relationship Between Lot Control and Chronic Defects]

Minamiyama: That's right...Kitaike, this lot is really one lot with the same manufacturing conditions, isn't it?

Kitaike: That's right. Headquarters will understand that lot control is the most important thing in the manufacturing industry.

Nishikawa: It's true that every time 4M changes, the lot number is changed, and since we chased this lot while we were there, nothing had changed...

Kitaike: Well, since the products are being produced continuously, there should be no need to mess with the equipment, and since the time zone is different from the shift time of 1st and 2nd shifts, there should be no particular change in people and methods. Hey...the only difference is the material. Shouldn't the coil of the material replaced here?

Nishikawa: Then, there seems to be a problem with the material in the morning.

Minamiyama: That's right. Nishikawa, do you have any idea why?

Nishikawa: Well, it's true that this product is a lot of 100,000 pieces, one coil is not enough, so we use two coils. However, I have not heard reports that there was an abnormality in the inspection when the material was put into stock, so what is the problem?

Kitaike: Oh my god, so the improvement is on-site and in-kind, isn't it? Now, ask the person in charge at the site to see if there is really a problem.

Minamiyama: Alright, let's go to the person in charge of purchasing materials immediately.

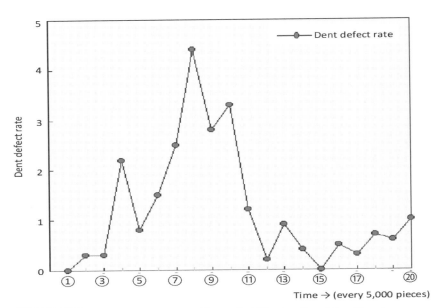

Figure 2.18 Time-Series Survey of Intra-Lot Fluctuations Line Graph

---

Coffee Break

(3) Eye Level During a Conversation

Question: Are you aware of the other person's eyes and your own eye level during conversations at the workplace during improvement activities?

►Example answer:

It's a good idea to match your own eye level to the other person's eye level as much as possible. For example, when the other person is sitting in a chair, bend down a little and look at the other person. If this is a relationship where one looks down on the other, the other person will cower. And if it is a relationship where one is looking up to the other, the other person will not take you seriously and it will be difficult to exert force.

When you talk to children, do you naturally talk at his/her eye level?

Everyone should have practiced how to communicate with others from their own experiences, but when it comes to improving the workplace, the mind can't relax and you forget this importance.

Step 5 Stratification for Factor Investigation (II) [Details]
[Search for Factors From the Causative System]

Already in the first half of Step 4, the strongest member to solve the problem has been selected. In the latter half of Step 4, you conducted a factor investigation of the result system from the viewpoint of phenomena. In Step 5, together with the strongest member, the cause system that causes the problem is searched in detail. If the defect rate is low, you should proceed with the investigation while considering the intentional occurrence of defects.

Step 5 can be roughly divided into the following three stages: early stage, middle stage, and final stage.

[Procedures 1 and 2] "Early Stage": "Investigation of the disparity between the ideal"

[Procedures 3 to 5] "Middle Stage" "Manufacturing factor (two out of the 5Ms) investigation"

[Procedures 6 and 7] "Final Stage": "Manufacturing factor (the last 2Ms) investigation"

In Step 5 "Early Stage," the strongest members will identify the gap between the ideal and the reality of where a product that should be a good product becomes defective by two procedures. Specifically, we will investigate the gap with reality in terms of both product functions and equipment functions. For the strongest team with all the key people in Step 4, the first thing that needs to be done in Step 5 to investigate the causative system of chronic defects is to confirm the ideal product functions and equipment functions.

The "Middle Stage" of Step 5 is divided into three procedures, which are the basic manufacturing factors 5Ms (material [Material], equipment [Machine], operator / worker [Man], manufacturing method [Method], measurement [Measurement].), "materials" and "equipment" are confirmed by the strongest members. Especially for "equipment," its "life" should be confirmed as well. In addition, the "measurement" of 5M has been confirmed in Step 3.

The "Final Stage" of Step 5 is divided into two procedures. Here, you will investigate the two Ms (operator / worker [Man], and manufacturing method [Method]) related to the strongest members themselves. If you haven't found the root cause of the failure by the Step 5 "Middle Stage," we recommend starting Step 5 "Final Stage" as a last resort.

In other words, this Step 5 "Final Stage" investigation is not necessary if a possible root cause has been found.

Each procedure in Step 5 is explained below.

[Procedure 1] Product Function Confirmation

In [Procedure 1], the strongest members will check how the product is used and the function of the components. In particular, we have to investigate what kind of condition of the processed part is ideal.

Procedure Points

(1) Identify the product functions.
- Check the usage of the product, specifications, drawings, meanings of the constituent parts, and forcibly manufacture quality defects. Anyway, let's carefully compare the actual product and find the gap.

(2) Check them with all the strongest teams of the company regarding the quality defect.
- It is a showcase of the skills of the strongest team.

[Procedure 2] Confirmation of Equipment Functions

After confirming the gap between the ideal product function and the current situation in [Procedure 1], the strongest member confirms the ideal movement of the manufacturing equipment (including molds, tools, etc.) and the role of each component of the equipment in [Procedure 2]. In particular, we will investigate what kind of condition is ideal for the part to be processed.

Procedure points

(1) Identify the equipment functions.

- Since the relationship between the generation process and its predecessor is clarified by the fluctuation survey between manufacturing processes, let's find the gap by carefully comparing the ideal state of the target manufacturing equipment with the actual equipment.

(2) Check with all the strongest teams of the company regarding the problem.

- It is a showcase of the skills of the strongest team.

[Procedure 3] Differences by Material / Parts Lot (Materials)

After confirming which points of the product or equipment should be focused on in [Procedure 1.2], proceed to the "Middle Stage" of Step 5. In [Procedure 3], we will investigate whether there is a difference in the trend of the defect to occur depending on the material lot.

Procedure Points

(1) Are there any differences between lots of materials and parts?

- A statistical method called a "control chart" that searches for changes between lots is effective. If these changes can be found within a certain investigation period, it will be easier to find the cause.

- It is important whether or not the intra-lot fluctuation survey has already been conducted. Depending on the sampling method, there are many cases in which changes due to different lots of materials and parts are overlooked, or conversely, misdiagnosis is made as if there were changes.

(2) The key is to discover the differences between strata.

- Is it clear what to investigate, such as different lots of materials/parts?

- It is a showcase of the skills of the strongest team.

(3) The sample is a representative of soup
- Is the sample representative of the group you want to compare?

Example) When you cook at home and taste the soup in the pot, you never evaluate only the top layer, but after stirring it, you taste a representative of the whole thing.

(4) Useful QC Methods / Analysis Tools
- Histograms, line graphs, bar graphs, control charts, etc.

(5) Particularly Useful QC Methods
- Figure 2.19 shows an application example to check if there is a difference in the lot distribution of materials / parts by using a histogram.

[Procedure 4] **Difference by Line (Machine-1)**

While [Procedure 3] confirmed the difference by material/parts lot, [Procedure 4] investigates whether there is a difference in the trend of the defect to occur due to the difference in the line or tool. Even if the equipment (including

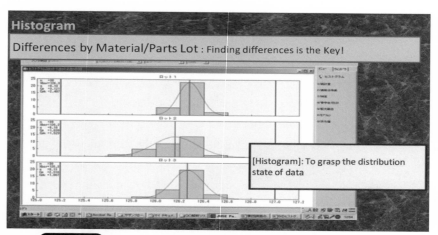

Figure 2.19 Histogram for Difference Survey by Material/Parts Lot

tools) is the same in design, the results will vary. It is necessary to suspect that they are similar but different.

(Procedure Points)

(1) Are there any differences by line?

- When manufacturing the same product on multiple manufacturing lines or molds, the lines actually are often slightly different even if the line design drawings are the same.
- Collect data carefully so that there are no flaws in the sampling method.

(2) Useful QC Methods / Analysis Tools

- Histograms, control charts, line graphs, bar graphs, etc.

(3) Particularly Useful QC Methods

- Figure 2.20 shows an application example to check if there is a difference in the distribution for each line by using a histogram.

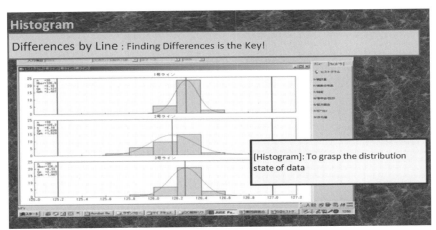

Histogram

Differences by Line : Finding Differences is the Key!

[Histogram]: To grasp the distribution state of data

Figure 2.20 Histogram for Line-Specific Difference Surveys]

[Procedure 5] Differences by Mold and Jiig (Machine-2)

After confirming the difference in the line and jig in [Procedure 4], in [Procedure 5], investigate whether there is a difference in the trend of the defect to occur due to the difference in the life of the mold or tool. Even if a defect item does not occur at the start of production, it may become a chronic defect during continuous production and become an "it is only natural if it occurs" situation. Check to see if there are any points that you haven't noticed before with the strongest members.

Procedure points

(1) Is there any difference between molds and jigs?
- When manufacturing the same product with multiple molds and jigs, the subtle differences between each figure and the assembled molds and jigs have been confirmed in the previous section. Therefore, in [Procedure 5], let's investigate the reliability such as the life of the manufactured mold and jig.

(2) Useful QC Methods / Analysis Tools
- Weibull analysis, photographs, etc.

(3) Particularly Useful QC Methods
- As shown in Figure 2.21, Weibull analysis is used to check if there is a difference in the life of each mold and jig (check [m value]: It is possible to identify the cause of initial failure or the cause of wear).

[Procedure 6] Are There Any Differences Between Workers?

From here, it will be the "Final Stage" of Step 5. In [Procedure 6], we will investigate whether there is a difference in the occurrence status of defects depending on the operator.

This investigation is conducted when the possible factors cannot be found

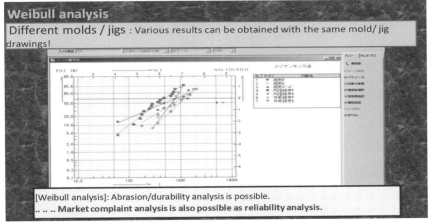

Weibull analysis

Different molds / jigs : Various results can be obtained with the same mold/ jig drawings!

[Weibull analysis]: Abrasion/durability analysis is possible.
.. .. .. Market complaint analysis is also possible as reliability analysis.

Figure 2.21 Weibull Analysis for Life Difference Investigation by Type and Jig

SOUTH-FLOW How to Promote Chronic Defect Elimination Activities by SOUTH-FLOW

in the whole previous investigations. Because they are the strongest members who have cooperated with various investigations so far, colleagues are willing to cooperate with this investigation. Also, around this time, it is not uncommon to request this survey from the strongest members who have gathered in search of personnel to improve it.

However, teamwork must be emphasized for future improvement. Therefore, this survey is one of the surveys that we would like to avoid if possible.

Procedure Points

(1) Is there any difference for each worker?

- When multiple people are involved in work on the same production line, the actual work is often slightly different even if the work standards are the same.

- Checking the quality of each worker's workmanship is the fastest way to verify this.

(2) Useful QC Methods / Analysis Tools

- Histograms, line graphs, bar graphs, photographs, videos, etc.

(3) Particularly Useful QC Methods

- Figure 2.22 shows an application example to check if there is a difference in the distribution by worker by using a histogram.

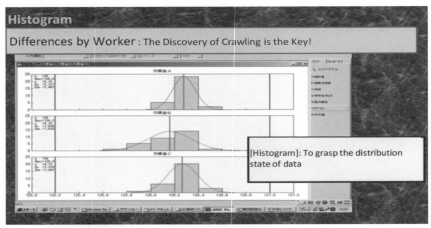

Figure 2.22 Histogram for Worker Difference Survey

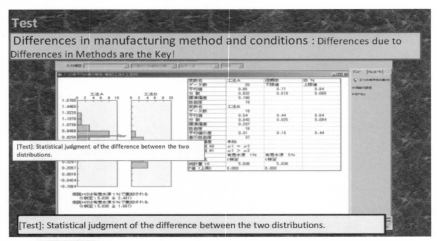

Figure 2.23 Certification for Difference Investigation by Manufacturing Method and Conditions

[Procedure 7] Differences by Manufacturing Method and Conditions (Method)

In [Procedure 7], we assume that the current manufacturing method or manufacturing conditions are the cause of the chronic failure, and search for other manufacturing methods or manufacturing conditions.

Similar to the worker survey, this survey is also conducted when the possible root cause cannot be found in the previous surveys. As mentioned above, they are the strongest members who have cooperated with various investigations so far, so if it is a chronic defect improvement, they would definitely cooperate to doubt the current manufacturing method or manufacturing conditions. In addition, around this time, it is not uncommon for the team to naturally come up with the search for the optimum construction method and confirmation of whether the current manufacturing conditions are optimal.

As we would like to reiterate, improvement that emphasizes teamwork is desired for future improvement, so the "Final" two surveys in Step 5 should be avoided if possible.

Procedure Points

(1) Are there any differences between manufacturing methods and conditions?
- Let's examine the manufacturing method and manufacturing conditions.
- For a final grasp of the current situation, let's investigate construction methods different from the relevant process such as other types and other companies. This is especially effective when no clear difference has been found in the current situation.

(2) Useful QC Methods / Analysis Tools
- Histograms, bar graphs, etc.

(3) Particularly Useful QC Methods
- Figure 2.23 shows an application example to check if there is a difference in the distribution by manufacturing method and conditions by using a histogram and a test.

[Improvement example 5] Product A Dent defect improvement-5
[Pursuit of the true cause by the strongest members]

Nishikawa: This Figure 2.24 shows a factor investigation of the causative system by SOUTH-FLOW.

Kitaike: Dent has been a chronic defect for a long time, it has the fate of a sheet metal shop, so to speak. From the beginning, I thought that the amount of waste for this Product A was only conspicuous due to the production volume, but this Southern thing, isn't it a pretty interesting guide?

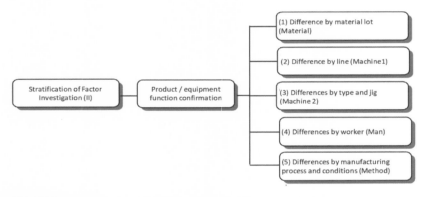

Figure 2.24 Stratification (II) Procedure (System Diagram) of Factor Investigation of SOUTH-FLOW

[Manufacturing Equipment Functions and Malfunctions]

Minamiyama: Kitaike said that the mechanism of the Dent generation is that foreign matter is pressed together with the material while being sandwiched between the punch and the die.

So, you can think of it as a Dent that looks like the picture in Figure 2.25.

Kitaike: That's true for most Dent that occurs in the press.

90

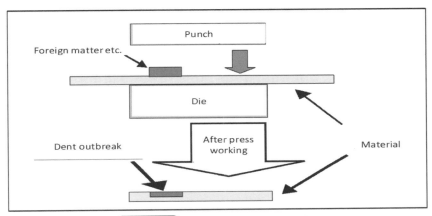

Figure 2.25 Image of Dent Generation

[A Slightly Different View of Material Lots]

The members who got a big hint from the analysis result of the phenomenon side went to the material receiving place and confirmed the following two points from the talk with the person in charge of acceptance inspection, Uemachi.

(1) No lot defects have been found in the inspection when the materials arrive.

(2) Due to the convenience of the material manufacturer, materials with different manufacturing conditions are received alternately as the same lot as shown in the image in Figure 2.26 (however, there is no functional quality problem of Product A due to this difference in conditions. This

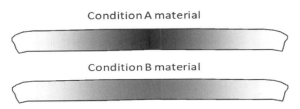

Figure 2.26 Image of Materials A and B With Different Manufacturing Conditions

has been proven in the past).

Minamiyama: Kitaike, does the difference in the manufacturing conditions of the materials have any effect on the pressed products?

Kitaike: Well...I don't know the details, but before, due to a manufacturer's mistake, materials with different manufacturing conditions were delivered and a large number of defects output, and all of them were discarded as lot outs. But just this time, it has been proven in the past that the conditions have nothing to do with the product functionality, so let's go to the site and see.

Members who went to the manufacturing site and checked the molds and products discovered new facts that they had not noticed before.

[What They Have overlooked So Far]

Minamiyama: How is it, Kitaike, did you see anything?

Kitaike: Yeah...I'm a little worried...maybe I grabbed some small evidence.

Kitaike, who picked up the product at the manufacturing site in a hurry, noticed something. Leader Minamiyama and the person in charge of Nishikawa are excited to see Kitaike carefully checking the product.

Nishikawa: What exactly did you find out? Please tell us quickly.

Minamiyama: Kitaike, what are you curious about?

Kitaike: Well, this is the product produced with the material of the first coil in the morning, and this is the product produced with the material of the second coil in the afternoon. Can you tell any difference by touching it? Nishikawa?

Nishikawa: Yeah...somehow, the product in the morning seems to have a slightly rougher cut surface.

Kitaike: Oh, it seems that even computer geeks have an eye for products. So do you know what that means?

Nishikawa: Um...because it is not cut well, the distortion at the time of cutting is large, the adjustment of the mold and the material is not good, and the maintenance of the mold is bad.

Minamiyama: I see, that's what it is...Kitaike, Nishikawa, you may be able to see the true identity of the defect, not small evidence. Let's analyze the cut surface by sampling tomorrow's products with different manufacturing conditions.

Kitaike: As expected, you know the way, Leader.

Nishikawa: What the hell is that? Don't be convinced by just those two.

The next day, Nishikawa collected samples as requested by the Leader, Minamiyama.

Nishikawa: Minamiyama, I got a sample. Condition A and condition B, products and scraps immediately after each material change.

Minamiyama: Thank you, this should surely clarify the relationship between the manufacturing conditions of the material and the roughness of the cut surface of the product.

Nishikawa: What on earth are you doing with this sample?

Minamiyama: I have heard that there is a machine that can take enlarged pictures in the test room of the Quality Assurance Department. If you magnify the cut surface of this sample and take a picture, you should be able to clearly see the difference between the material of condition A and the material of condition B. I thought that the cause might be revealed if I had the material properties investigated, so I contacted the person in charge of analysis of the Quality Assurance Department, Shimoda, and made a reservation yesterday. Let's visit him immediately.

[Big Harvest From Small Discoveries]

As a result of taking a magnified photograph of the cut surface of the sample at the Quality Assurance Department, important facts emerged from the photograph.

Minamiyama: As you can see, Kitaike, it's true that most of the Dent defects occur when burrs and debris during pressing get into the mold, right? What do you think of the photo in Figure 2.27?

Kitaike: Yeah, it's the first time I've seen a photo like this, so I can't say anything...however, as you can see, the material of condition A seems to have a much rougher cut surface.

Minamiyama: What do you think, Nishikawa?

Nishikawa: Yes, I feel the same as Kitaike, but what seems to stick to the cross section of condition A is the residue and burrs generated during cutting, right?

Kitaike: In other words, when it falls into the mold and is pressed while being sandwiched between the materials, Dent defects occur...is that what it means?

[Deciphering Facts (Graphs) With the Strongest Members]

Nishikawa: That's why in the morning when the material of condition A is used, the number of defects increases in a blink of an eye. But why is the number of defects showing a sharp decrease in the afternoon?

Kitaike: That's because they changed the material to the one of condition B, didn't they? You didn't even know that, so you just make defects.

Nishikawa: Yeah, that's right...that means...well, that's right.

Kitaike: What did you convince yourself of?

Nishikawa: Yes, the defect transition graph in Figure 2.28 shows that the defect occurrence situation that we regarded as a series of flows until yesterday included two elements, condition A material and condition B material. In other words, it is appropriate to think of the data (10) and (11) in Figure 2.28 as separate graphs before and after the material change.

Minamiyama: I see, that's exactly what Nishikawa said. As shown in the graph of Figure 2.29, defects have increased sharply immediately after the start of production. Conditions (1) to (10) are materials for condition A, and condition B shows a slight increase in the overall defect occurrence rate. There is no connection between the materials and the two.

Kitaike: Oh, it looks like you can do it if you try, Nishikawa. I guess you're not just a computer geek. Speaking of which, even before this product was launched, defective Dent had been a problem as a chronic defect, but everyone thought that this was unavoidable and no one tried to do anything about it. If we think about countermeasures from this result, we may be able to solve them altogether.

Minamiyama: That's right. But first, let's consider the countermeasures for condition A material, which causes a large amount of defects. If you think simply, it is necessary to take measures for the mold that can handle both conditions A and B, but are there any good ways?

Kitaike: Well, I think there are many ways to do it...I think it would be better to open this to the mold factory guys, so I'll run and talk to them. With all this evidence, our professional mold members should be able to do it.

Minamiyama: Yeah, thank you.

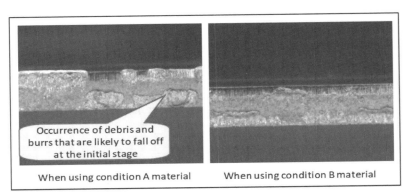

Occurrence of debris and burrs that are likely to fall off at the initial stage

When using condition A material   When using condition B material

Figure 2.27 Photograph of a Cut Surface When Using Different Manufacturing Conditions

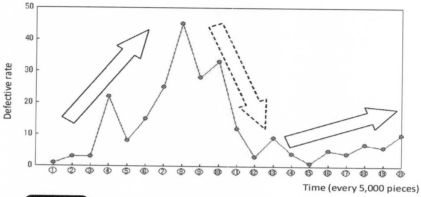

Figure 2.28 Time-Series Survey of Intra-Lot fFuctuations Line Graph

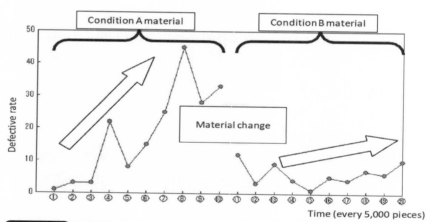

Figure 2.29 Time-Series Survey of Intra-Lot Fluctuations and Materials Used Line Graph

─────── **Tea Time** ───────

(3) Is being asked to "Remeasure it!" statistical sense?

As long as we are in the manufacturing industry, we always evaluate whether the finished product meets the drawing standards. No one would be able to accept it if a product they made was concluded to be out of specification, specially for workers who are confident in their skills. Have you ever heard an exchange that includes "Could you measure it again?" This is exactly a complaint about the evaluation result.

In fact, the person who makes such complaints can be said to have statistical sense.

-The Law of Large Numbers-

The law of large numbers is when n is made infinitely large, the variation of the X (x bar) becomes infinitely small. The following formula holds.

$V (X \text{ bar}) = \sigma^2 / n$ [V: Variance]

"Can you measure it again?" means that someone cannot trust whether the evaluation is correct in one measurement, so please re-evaluate it. In other words, they want to confirm the true value of the product by measuring the same thing several times and make a correct evaluation.

The important thing here is to think of the average value, which is measured several times, as the true value of the product. Don't forget that it is nonsense to measure it several times and evaluate it when you get the value you want.

*It is necessary to determine whether there is any problem with the conventional measuring machines in response to the customer's request for accuracy improvement.

Step 6 Organization of Factors
[Organize Factors with the Strongest Members Using Facts as Hints]

After investigating the cause of the causative system in Step 5, in Step 6, organize the factors together with the key people who have convened so far. There is only one procedure in Step 6. Sort out the factors using the facts as hints in order to improve chronic defects by grasped facts. Avoid arranging the factors only on the desk by using K・K・D・H (KAN intuition, KEIKEN experience, DOKYO courage, and HATTARI bluntness).

The procedure in Step 6 is explained below.

[Procedure 1] **Arrangement of the Relationship Between Defect Characteristics and Factors**

The factors found from the result and cause system so far are organized as linguistic data. Based on the survey results so far, we will organize it specifically to a level where we can experiment in the next step with the strongest members that were gathered by key people. Depending on how this is organized, whether or not the boss can get an understanding of the budget and man-hours to be spent on the next analysis experiment can be decided.

Procedure Points

(1) Group the listed factors and organize them visually.
- In addition to the knowledge gained by grasping the current situation so far, including experience and intuition, let's repeatedly ask why, and summarize it as linguistic data.
- Let's summarize it as linguistic data, such as whether it is a factor that influences the average of characteristic values or a factor that influences the variation.

(2) If the cause is unknown and it is difficult to sort out, reconfirm the current situation.
- Are there any unconfirmed items in the current situation?
- For defects with a low incidence ratio, it is effective to conduct experi-

ments such as forcibly creating defects. Which factor is likely to be used to produce a defective product?

(3) Useful QC Methods / Analysis Tools
- Ishikawa diagrams, system diagrams, etc.

(4) Particularly Useful QC Methods
- Figure 2.30 shows an application example of organizing language data based on the "hints" and "experiences" confirmed in the scenarios so far, using a system diagram.

[Improvement Example 6] Product A Dent Defect Improvement-6
[Efficient Factor Arrangement Based on Facts]

Minamiyama: In order to sort things out, let's sort out the causes of the defects obtained in the survey so far. "SOUTH-FLOW" positions the organization of factors as important in order to promote the understanding of the parties concerned and lead to efficient factor analysis.

Nishikawa: That's right. Here, let's sort out "Why do Dent failures occur?" using a factor-seeking system diagram. First of all, among the five

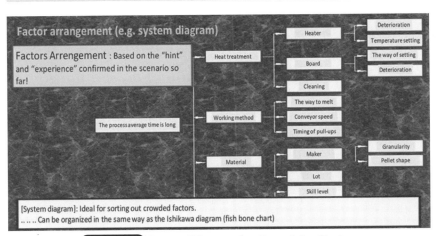

Figure 2.30 System Diagram for Organizing Factors

processes related to this product: press, barrel, heat treatment, surface treatment, and visual inspection, the press process had an overwhelmingly high defect rate.

Minamiyama: Yes, and as a result of conducting a survey by dividing the pressing process into materials, dies, equipment, and handling, we narrowed down that the most important point of defect occurrence were dies.

Chief Kitaike ran to them with a smile.

Minamiyama: Oh, Kitaike is back.

Kitaike: Minamiyana, the leader said okay. They say it is possible to handle both conditions A and B materials with a simpler repair than we expected. I know they are specialists in precision press products. Uchida, the chief of the mold section, will take care of everything from repair plans for improvement to arrangements for the mold factory, and when today's production is finished, he will be able to handle urgent mold repairs for night shifts.

Nishikawa: You did it! It seems we can clean up this mess.

[Mold Wear and Chronic Defects]

Minamiyama: Kitaike, I'm just working with Nishikawa to organize the factors as shown in Figure 2.31. In this way, I understand that the deterioration of the cross section in the mold, not "R molding" but especially "outer circumference removal," leads to defective Dent. Kitaike, is there anything that can be considered as a factor that leads to deterioration of the cut surface without the outer circumference?

Kitaike: Well, the deterioration of the cut surface means that the mold deteriorates, so the first thing that can be considered is the wear of the mold.

Minamiyama: Indeed, there is also a confirmation of mold wear in Step 5-(3) of this SOUTH-FLOW. So the other day, I had Nishikawa get a sample to check the cut surface.

Nishikawa: Hey, I left the envelope with the photo I received last night in my bag. Hmm, where was it, oh, here it is. Figure 2.32 is a photograph of the sample. As requested, I had them take pictures of samples immediately after maintenance and after production of 250,000 pieces from the site.

Kitaike: You did great, Nishikawa! How good is our teamwork? Well, if I remember correctly, at the end of the year when the in-house QC circle was announced, the mold-related circle announced an improvement in material or something. After all, the increasing trend of that defect is that it becomes the system diagram of Figure 2.33. However, I wonder if this Southern thing has something like that in it.

Minamiyama: It's a hit. There is a clear difference in the cut surface. Mr. Kitaike, is there anything else? We're relying on the skill of the workers, or something like the so-called "world of craftsmen…"

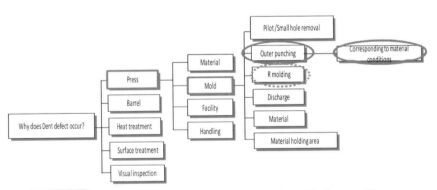

Figure 2.31 Arrangement of Causes of Dent Defects (1) System Diagram

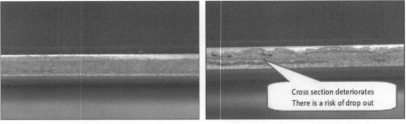

Immediately after maintenance   After the production of 250,000 piecesb

☆ One of the reasons why the Dent defect rate gradually increases in the press lot is that the residue (burr) falls off due to the deterioration of the punched section.

Figure 2.32 Photograph of the Cross-Section Comparison for Tool Life Survey

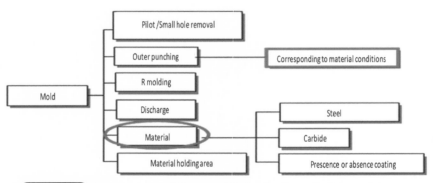

Figure 2.33 Arrangement of Causes of Defective Dent (2) System Diagram

[Optimal Conditions for Chronic Defects]

Kitaike: Other factors that, huh...by the way, these are "Mold clearance A" and "Mold size B" aren't they? These two influence each other, and it is quite difficult for even a veteran to set the best conditions. If this is also added to the system diagram, it will look like Figure 2.34.

Minamiyama: I see, that's what I got.

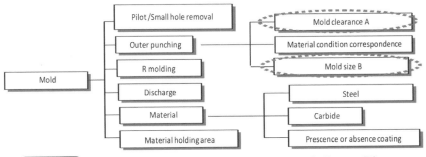

Figure 2.34 **Arrangement of Causes of Defective Dent (3) System Diagram**

SOUTH-FLOW How to Promote Chronic Defect
Elimination Activities by SOUTH-FLOW

Coffee Break

(4) Personal Space

▶ **Question:**

How much do you think about the distance from the other person (actual standing or sitting position) during conversation?

▶ **Example answer:**

When asking for a status survey in any workplace, it is generally appropriate to be in reach of each other.

Each person has a comfortable distance from the other person, for example, lovers do not feel uncomfortable even if they touch each other. But when you, the improvement promoter, try to talk at a short distance, smooth communication is difficult.

On the other hand, if you try to listen at a distance where if you reach out and your hands can't touch, you will feel awkward, which also makes communication difficult. If you always keep in mind the distance to the other person, communication for improvement will be easier than you imagine even if it is a new workplace.

Step 7 Experiment for Factor Analysis (I)
[Searching for "Contribution Factors" From "Possible Factors"]

It is very difficult to think about and implement countermeasures for all of the factors listed in Step 6. Therefore, with SOUTH-FLOW, we will experimentally confirm how much the "possible factors" mentioned are affecting the causes of chronic defects. There is only one procedure in Step 7, and here we will check the impact of many "possible factors" while considering the entanglement of the factors listed. For details of the L8 orthogonal array experiment in the design of experiments, refer to Chapter 3, Section 3.3.

The procedure in Step 7 is explained below.

[Procedure 1] **Experiment by Allocating Possible Factors to Two Levels**

Narrow down the root cause from the possible factors listed in Step 6 thorough experiments. Here, we will make the best use of design of experiments to carry out efficient analysis experiments. Especially in the case of chronic defects caused by complicated mechanisms, there are many cases in which the factors have been intertwined and have an effect, and have not been resolved until now. In such cases, the design of experiments may be used to solve the problem at once. In SOUTH-FLOW, if you want to make sure that you can experiment with up to two "possible factors" in the previous Step 6 or several factors including variations at once, skip Step 7 and go to the next step. Go directly to Step 8.

Procedure Points

(1) If there are up to 6 factors, perform an L8 orthogonal experiment.
- Check the influence of factors that contribute to the average value and defect rate.
- If there are few "influential factors," consider "variation" from the beginning and conduct an efficient experiment.
- f there are "many possible factors" as the cause of quality defects, it is desirable to confirm the degree of influence of the factors at a lower cost.

- Knowing the entanglement (interaction) of factors encourages the next idea (if the factors are not entangled, are they solved one by one?).

(2) If the number of factors are from seven to 14, perform an L16 orthogonal experiment.

- If "many (seven or more) possible factors" are the cause of quality defects, it is desirable to first confirm the degree of influence of the factors at a lower cost.

(3) Be careful about side effects.

- Check the effect on other quality characteristics. In particular, work closely with your boss and quality assurance personnel.

(4) Useful QC Methods / Analysis Tools

- Orthogonal array experiments

(5) Particularly Useful QC Methods

- Figure 2.35 shows an application example of the factor effect plot diagram obtained from the orthogonal experiment of the two-level system in order to confirm the influence degree and entanglement of the factors.

[Improvement Example 7] Product A Dent Defect Improvement-7
[Efficient "Possible Factor" Survey]

Among the failure factors that have been sorted out, the members cited two factors as "possible factors": "mold clearance" and "mold size B," as can be seen from the system diagram in Figure 2.36 below.

Nishikawa: Minamiyama, the "possible factors" for removing the outer circumference of the mold are the two factors that rely on the "skills" of the mold correctors.

Minamiyama: That's right. If more factors can be considered, there is an efficient design of experiments called orthogonal experiment, but this time we only need to investigate two factors, so we decided to skip to "Step 8" following SOUTH-FLOW.

Nishikawa: I remember it was written somewhere that "Step 7" is useful when there are many "possible factors."

Minamiyama: Nishikawa, you're really good at remembering. You're thinking, "I'm sure I'll be a problem solver soon."

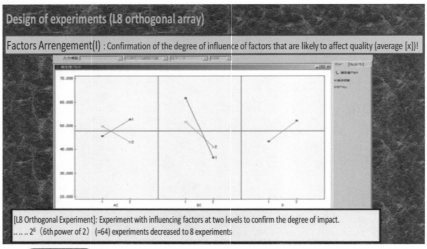

Figure 2.35 Factor Effect Plot for Factor Analysis L8 Orthogonal Array experiment

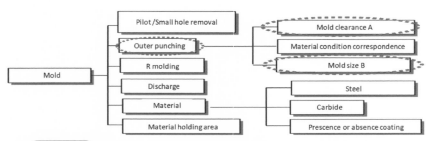

Figure 2.36 Arrangement of Causes of Dent Defects (4) System Diagram

─────────── **Tea Time** ───────────

(4) Is saying, "Let me measure it!" statistical sense?

As long as we are in the manufacturing industry, we always evaluate whether the finished product meets the drawing standards. No one would be able to accept it if a product they made was concluded to be out of specification. Especially for workers who are confident in their skills.

Have you ever heard an exchange that includes "Let me measure it!"? This is exactly a complaint about the evaluation result.

In fact, the person who makes such complaints can be said to have statistical sense.

-Additivity of Variance-

The additivity of variance means that in the basic form where X and Y are independent and the constant is 1, the variance (V) relationship as shown in the following equation holds.

$$V(X+Y)=V(X)+V(Y) \quad [V: Variance]$$

"Would you like me to measure it?" means that they suspect that the measured value was judged to be defective due to the difference in measurement by humans, so they would like you to change the measurer and re-evaluate.

What is important here is "repeatability of measurement: what is the difference in reading value when repeatedly measuring the same object?" and "reproducibility of measurement: reading value when measuring by changing the measurer." It is necessary to confirm "What is the difference between the two?" and evaluate the product correctly. When someone else gives you a value they want to measure, we don't evaluate it with that value.

The rule is that the standard deviation (statistically, "standard deviation: s [Standard Deviation]" is the correct answer), commonly referred to as "σ (sigma)", cannot be added, but "σ^2 (s^2)" can be added.

Below are some useful and simple formulas for eliminating chronic defects.

① $\sigma_A^2 + \sigma_B^2 = \sigma_M^2$ [Variation of general measurement as a whole]

σ_A^2: Variation due to "measurement repeatability"

σ_B^2: Variation due to "measurement reproducibility"

σ_M^2: Variability of the entire measurement

→The overall variability of the measurement is the sum of the variability of measurement repeatability and reproducibility.

*A stable measurement method enhances the reliability of evaluation no matter who measures it at any time.

② $\sigma_M^2 + \sigma_{LN}^2 + \sigma_{LK}^2 = \sigma_T^2$ [General variation in overall manufacturing quality]

σ_M^2: Variability of the entire measurement

σ_{LN}^2: Variation within production lots

σ_{LK}^2: Variation between production lots

σ_T^2: Variation in overall manufacturing quality

→The overall variation in manufacturing quality is the sum of the overall measurement variation and the variation within and between production lots.

*In manufacturing, there are variations within and between lots, but keep in mind that it is difficult to evaluate the manufacturing quality that should be evaluated due to the large variation in the entire measurement.

Step 8 Experiment for Factor Analysis (II)
[Searching for Optimal Conditions for "Contribution Factors"]

After narrowing down the "influenced (contribution) factors" out of the many "possible factors" in Step 7, the next step is to conduct an impact survey including the variability in Step 8. There is only one procedure in Step 8, and here, we confirm the optimal conditions for influential factors (contributive factors) using design of experiments such as factor placement (single, binary, multiple) experiments and orthogonal array experiments. See Chapter 3, Section 3.3 for more information on commonly used design of experiments.

The procedure in Step 8 is explained below.

[Procedure 1]	Experiment by Allocating Contributing Factors to Three Levels

Further experiments will narrow down the root causes of the contributing factors narrowed down in the previous step. Here, we will make the best use of design of experiments to carry out efficient analysis experiments as well. In particular, when selecting the optimal conditions, also investigate "variation."

Procedure Points

(1) Check the degree of influence of factors that contribute to variability.
- Let's carry out an efficient experiment considering the optimal conditions of "influencing factors" and "variation."
- If there are few "influential factors," consider "variation" from the beginning and conduct an efficient experiment.

(2) Be careful about side effects.
- Check the effect on other quality characteristics. In particular, work closely with your boss and quality assurance personnel.
- If other characteristics could be affected, choose a well-balanced optimal condition.

(3) Useful QC Methods / Analysis Tools
- Orthogonal array experiments, placement experiments

(4) Particularly Useful QC Methods

- Figure 2.37 shows an application example of the factor effect plot diagram obtained from the multidimensional placement experiment (combination of level 4 and level 3) in order to confirm the degree of influence and entanglement of factors.

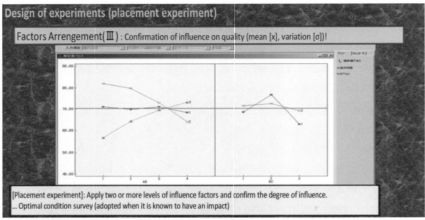

Figure 2.37 Factor Effect Plot Diagram for Factor Analysis Factor Placement Experiment

[Improvement Example 8] Product A Dent Defect Improvement-8
[Understanding the World of Craftsmanship]

Among the failure factors that have been sorted out, the improvement team who narrowed down the true causes to "mold clearance A" and "mold size B" obtained the consent of the general manager of Tokai Headquarters and obtained the cooperation of the mold factory and the technical department. They have reached the point of conducting "experiments for factor analysis" utilizing analysis of variance and design of experiments.

Kitaike: Now, let's unravel the "world of craftsmen" that left behind the "arms" of the mold correctors. So what should I do, leader?

Minamiyama: I'm also excited. First of all, there are two "possible factors," "mold clearance A" and "mold size B," so I would like to compare about three ways for each.

Kitaike: Uchida, the chief of the mold section, gave me a proposal for an experiment last minute without breaking the mold, saying that I could prepare tools like this.

Minamiyama: To summarize the plan, I wonder if this experiment will look like Figure 2.38. I would like to repeat the same experiment twice, considering the comparison by the defective rate of Dent.

Nishikawa: I wonder if that might be the case, and Chief Uchida has okayed giving me two days so that I can experiment this weekend.

Kitaike: When did you become such a nifty guy?

Factor	Factor content (Possible factors)	Level 1	Level 2	Level 3
Factor A	Mold clearance A	small	mid	large
Factor B	Mold size B	small	mid	large

(Repeated experiments twice at each level)

Figure 2.38 **Experimental Design (Factor-Level Table) Matrix for Factor Analysis**

[Analysis of Factor Analysis Experiment]

The three members were able to carry out the experiment on the weekend safely under the cooperation of the mold corrector, including Chief Uchida of the mold section. In addition, certified visual inspectors have also obtained inspection results under each condition.

Minamiyama: Thanks to you, we got the result shown in Figure 2.39.

Nishiyama: This is an analysis of variance, isn't it? This * mark is...

Kitaike: When did you learn those words? I'm so surprised.

Nishikawa: Well, I only know that the * mark has an effect.

Minamiyama: It's amazing just to know that, Nishikawa. Then why don't
you both try something like Figure 2.40?

Kitaike: Hey leader, I know this. If "mold clearance A" is set to A3, the
defective rate will decrease. Compared to this, I don't want to see
the table of numbers called "dispersion or something". However, it is
strangely intertwined.

Minamiyama: Yes, it seems best to choose A3. Also, this entanglement
may have been the "world of craftsmen." This is a factor-effect plot
diagram, which is a diagram of the factors marked with * in that table.
Therefore, if you don't have an analysis of variance table, you won't
know which factor is marked with *, so that table is essential.

Factor	Sum of squares	Flexibility	Dispersion	Dispersion ratio	Test
Factor A	0.03445	2	0.01722	98.615	**
Factor B	0.03298	2	0.01649	94.417	**
Factor A × B	0.01613	4	0.00403	23.086	**
Error e	0.00157	9	0.00017	–	
Total	0.08513	17	–	–	

Figure 2.39 Experimental Results for Factor Analysis Analysis of Variance Table

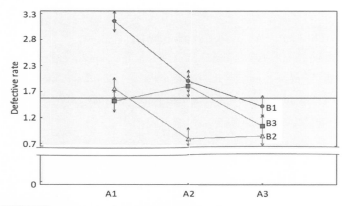

Figure 2.40 Experimental Results for Factor Analysis Factor Effect Plot

Step 9 Implementation of Measures
[Introduction of Optimal Conditions for Mass Production]

In Step 9, we will consider and systematically implement countermeasures to be introduced in mass production, while considering the degree of influence of the "contribution factors" obtained in Step 8.

The procedure in Step 9 can be divided into two parts.

[Procedure 1] Based on the information obtained until Step 8, consider specific countermeasures to be introduced into mass production and formulate an implementation plan.

[Procedure 2] At the time of planning, when introducing mass production of countermeasures, we will implement countermeasures after clarifying what kind of evaluation and notification application to customers are required.

Each procedure in Step 9 is explained below.

[Procedure 1]	Consider concrete measures and formulate implementation plans

From the information obtained up to Step 8, we will formulate the measures that should be incorporated into mass production necessary to achieve the purpose. Avoid unnecessarily incorporating all measures just because it is better or safer to take measures against the factors. A company's countermeasures against chronic defects require control over the root cause of the source, but it should not mean that it is sufficient to spend money by taking all measures. If multiple root causes are found, you should work with key people, including your boss, on measures to eliminate chronic defects, taking into account the impact of causing the defect.

Procedure Points

(1) Perform confirmation experiments multiple times (larger sample sizes).

- Before introducing mass production, thoroughly identify the problems that are expected or concerns. In particular, work closely with your boss and quality assurance personnel.

(2) Not all measures are required.

- Always be aware of costs and make decisions by referring to the impact evaluation of factors. The purpose is to eliminate chronic defects, so avoid excessive measures just because you are worried.

(3) Implement convincing measures together with the parties concerned.

- It will be a good improvement activity only if it is recognized by the people concerned.

(4) Useful QC Methods / Analysis Tools

- Matrix method, etc.

[Procedure 2] Implementation of Mass Production Measures

Once the mass production measures are decided, we will implement them as planned. Since this is a major manufacturing change, we will make sure to keep records before and after the countermeasures, and if the customer needs to apply them in advance, we will implement the plan according to the rules.

Procedure Points

(1) Implement measures according to the implementation plan.

- Describe the measures you have taken in detail so that you can see the difference before and after the measures.

(2) Countermeasure lots are a major change in production.

- Keep a record of countermeasure lots and don't forget to check lots before and after countermeasures (first product management: traceability).

(3) Have you completed the process change notification to the customer?

- Depending on the content of the measures, if a process change notification is required, the process change rules must be observed.

(4) Useful QC Methods / Analysis Tools

• Matrix method, etc.

[Improvement Example 9] Product A Dent Defect Improvement-9
[Implementation of measures against defective Dent]

Nishikawa: We could find various countermeasures.

Minamiyama: That's right. Thanks to everyone's cooperation.

Kitaike: Well, I don't think there will be such a countermeasure for that chronically defective Dent defect.

Nishikawa: In the end, there were three major points: (1) mold modification as a response to different material conditions, (2) tool material change as a tool wear countermeasure, and (3) discovery and education of an appropriate mold correction method.

Minamiyama: That's right, I would like to see the results after the introduction of mass production as soon as possible.

Nishikawa: A person from the Quality Assurance Department also said that this improvement is not a change that needs to be notified to customers, and I am really looking forward to the internal results after the measures.

Step 10 Effect Confirmation

[Confirmation of Effects After Introduction of Mass Production Measures]

Check items related to quality after countermeasures are also organized in Step 9, and the effectiveness of countermeasures is confirmed in Step 10. The procedure in Step 10 can be divided into three parts.

[Procedure 1] Check the effect,

[Procedure 2] Compare it with the target value.

[Procedure 3] Check the ripple effect as well.

We will check all of these in-house without causing any inconvenience to our customers.

Each procedure in Step 10 is explained below.

[Procedure 1] Confirmation of Improvement Effect

Clarify the tangible effect by giving specific numerical values for the improvement effect. Check them with the key people for side effects on other quality characteristics.

Procedure Points

(1) Implement measures according to the implementation plan.
- Set an appropriate confirmation period in consideration of the "problem occurrence frequency" and "occurrence cycle," and confirm the effect.

(2) Be careful about side effects.
- Check the effect on other quality characteristics. In particular, work closely with your boss and quality assurance personnel.

(3) Useful QC Methods / Analysis Tools
- Pareto charts, histograms, control charts, line graphs, bar graphs, pie charts, band graphs, radar charts, etc.

(4) Particularly Useful QC Methods
- Figure 2.41 shows an application example of the histogram used to confirm the effect by comparing the distribution before and after the improvement.

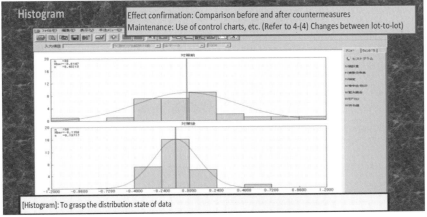

Histogram

Effect confirmation: Comparison before and after countermeasures
Maintenance: Use of control charts, etc. (Refer to 4-(4) Changes between lot-to-lot)

[Histogram]: To grasp the distribution state of data

Figure 2.41 Histogram for Confirming the Effect

[Procedure 2] Comparison With the Target Value

After confirming the improvement effect, compare it with the initially set target value. If the target is not achieved, we will review again whether there are any omissions in the survey, the content of the survey, and the analysis results.

⸻ **Procedure Points** ⸻

(1) Measure the effect after countermeasures and compare it with the target value.
  - Let's compare the measures before and after and with the target value.
  - Let's also check the tangible effect (economic effect).
(2) Useful QC Methods / Analysis Tools
  - Pareto charts, histograms, control charts, line graphs, bar graphs, pie charts, band graphs, radar charts, etc.

[Procedure 3] Grasp the Results and Confirm the SpilloverEeffect

The side effects are confirmed in [Procedure 1]. Here, we will compare the improvement effects other than the quality characteristics before and after the improvement even if the initial target is not set. Also, the intangible effect is a very important effect for companies, so don't forget to check it.

When searching for the optimal conditions in the experiment for factor analysis (II) [Searching for optimal conditions for "contribution factors"] in Step 8, if the quality evaluation characteristics are raised in addition to the defect, don't forget to check the spillover effect confirmation item during mass production.

> ### Procedure Points
>
> (1) Summarize the spillover effect.
> - Let's grasp tangible direct results (spillover effect) other than the target characteristic value.
> (2) Also check the intangible effect.
> - Are there any intangible effects that accompany the direct effects?
>   For example, it became a technical asset because it was possible to visualize that multiple factors are intertwined in the mechanism of occurrence of chronic defects.
> (3) Useful QC Methods / Analysis Tools
> - Pareto charts, histograms, control charts, line graphs, bar graphs, pie charts, band graphs, radar charts, etc.

[Improvement Example 10] Product A Dent Defect Improvement-10
[Review of SOUTH-FLOW (1): Phenomenon Survey]

Minamiyama: Let's review "SOUTH-FLOW" again here, because it will definitely be applicable to other improvements in the future.

Nishikawa: Well, first of all, of the six items in Step 4 proposed in Figure

2.42, "Stratification of Factor Surveys (I)", (1) trends by product, (2) fluctuations between manufacturing processes, and (3) trend of intra-lot fluctuation that we surveyed, we found a very interesting fact. The defective rate of Dacon defects tends to rise, fall, and rise slightly over time during the day, especially in the intra-lot fluctuation of (3).

Minamiyama: Yeah, when I found out the intra-lot fluctuations of this Dent defect, I switched from a data-based survey to a site-based survey, but since we were able to confirm the direction of improvement here, I don't have to investigate the remaining three items.

Nishikawa: I see, these six items are just the points to effectively grasp the current situation from the phenomenon side, so it is not always necessary to investigate all the items.

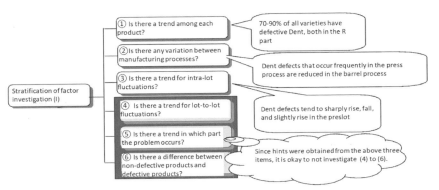

Figure 2.42 Stratification for Factor Investigation(I) and Comments System Diagram

[Review of SOUTH-FLOW (2): Detailed survey]

Minamiyama: Next, we entered into the "Strategies of Factor Investigation (II)" of "SOUTH-FLOW." What kind of scenario was this?

Nishikawa: Well, in this "stratification of factor investigation (II)," the image is to pursue the cause of the defect from the 4M perspective including equipment and materials, specifically, product functions, equipment functions, etc. There was an item about the change of 4M.

Nishikawa: There was a difference found in the incidence of defective Dent even though they were in the same lot, which led to the two types of manufacturing conditions for the material, and we observed the cut surface. Then, the mold led to measures corresponding to both manufacturing conditions of the material, and the effect was first achieved.

Minamiyama: And as shown in Figure 2.43, the next point was to pursue the cause of chronic Dent defects regardless of the manufacturing conditions of the material.

Nishikawa: Well, we wondered why burrs and debris that cause defective Dent occur even if the material is changed to the one with condition B?

Minamiyama: So, by stratifying "(3) Differences between molds and jigs" regarding mold wear, it was possible to visualize how the mold conditions are changing by observing the cut surface.

Nishikawa: When I showed it to the people involved in the mold and explained it, they immediately implemented the request for material change.

Minamiyama: Certainly, that was like the horizontal deployment of measures that QC circles in other departments worked on to improve the durability of tools that are completely different from Dent.

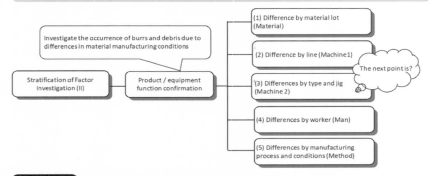

Figure 2.43 Stratification for Factor Investigation(ll) and Comments System

[Review of SOUTH-FLOW (3): Searching for Optimal Conditions]

Nishikawa: The last challenge was to the world of craftsmen who were surprised by the people involved in the mold and the workers.

Minamiyama: It was an interesting discovery that the dimensions and clearance

Kitaike just got there, he was holding some materials with great care.

Nishikawa: What's wrong, Kitaike?

Kitaike: I got this defect rate transition table from the site, thank you very much.

Minamiyama: Is it something wrong?

Kitaike: That's right, the girl Nishikawa made cry about the Ishikawa diagram, yes, Nakamura brought me the transition table of Figure 2.44, but she said, "Thanks to the project team, we operators were falsely accused, but it has been cleared up."

Nishikawa: I didn't make her cry.

Minamiyama: Well, Nishikawa, no one really thinks you cried. So, Kitaike, how much did the defects decrease?

Kitaike: The result is that the defect rate, which was 1.5%, is about 0.15%.

Nishikawa: Well, isn't the goal of "one-tenth the quality of orders of magnitude"?

Minamiyama: Anyway, I think that the Tokai Headquarters will be pleased with this, but what makes me even more happy is that the people at the site are concerned about the effect of the improvement, and they are pleased with the good results together.

Kitaike: With this improvement, we shared various discoveries with everyone in the field, which may be the reason why we were able to maintain a good atmosphere.

Minamiyama: After all, it's thanks to the two people involved in the field, as well as people involved in molds and quality assurance.

Nishikawa: Looking back, there were three of us in the project team, but a lot of colleagues cooperated. Uemachi, who is in charge of acceptance inspection, Shimoda, who is in charge of analysis in the Quality Assurance Department, Uchida, who is in charge of the mold section, Nakamura, who is in charge of manufacturing, and at least four people are involved besides us...except for Nakamura, who belongs to the same section to me, there are only people who worked together for the first time while working at the same factory. And in fact, there are more than a dozen people such as appearance certification inspectors and mold correctors. They were all very supportive.

Minamiyama: Nishikawa, for me, you, Kitaike, and all of them, it was their first time working. However, we needed the cooperation of everyone to improve this Dent defect. The strongest members were gathered exactly for solving Dent defects.

Kitaike: It's true that the leader didn't know the strongest members that gathered together...is this thanks to the Southern thing?

Nishikawa: It sounds a little strange, but it was thanks to this SOUTH-FLOW.

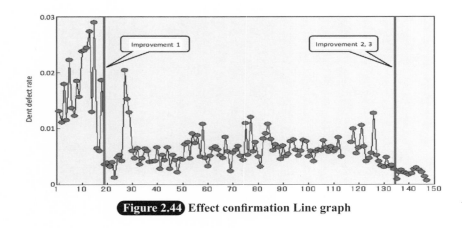

Figure 2.44 Effect confirmation Line graph

─────── **Coffee Break** ───────

(5) Smiles and Frowns

▶**Question:**

Are you aware of your facial expressions during conversations when you start improvement activities in your new workplace?

▶**Example answer:**

Let's talk with a friendly smile as much as possible. If you want to avoid a good relationship, you just have to look awkward.

From the point of view of people in the new workplace, you are an outsider, so if you do not greet them with a smile, it will take a long time to get to know each other. A friendly smile is the best weapon if you want to introduce smooth improvements.

With the addition of humor, it won't take long to get to know each other. In order to create a good atmosphere of improvement and increase the number of people who can improve, it is important to remember that your own mind is the key.

SOUTH-FLOW How to Promote Chronic Defect
Elimination Activities by SOUTH-FLOW

Step 11 Standardization
[Succession of Improvement Measures to the Next Generation]

After confirming the improvement effect in Step 10, standardization is performed in the final step to ensure that it will be succeeded to the next generation. The procedure in Step 11 is divided into three parts as follows.

[Procedure 1] Standardize the management method for maintenance and improvement.

[Procedure 2] Check the countermeasures with everyone at the site,

[Procedure 3] Confirm the challenges to the improvement presentation and the remaining problems.

Each procedure in Step 11 is explained below.

[Procedure 1] Determination of management method and establishment/revision of standards

Determine and standardize the management method without omission so that the effects confirmed in Step 10 will be passed on to the next generation. Don't forget to decide how to manage each of them even if there are multiple root causes. If horizontal deployment to other products is possible, deploy with related parties.

Procedure Points

(1) Create a curb to maintain the effect.
- Let's consider a curbing measure with 5W1H.
- Make full use of digital cameras and videos, incorporate them into check sheets and work standards, and use them as educational materials.

(2) A "memorial tree" that will be passed down to the next generation.
- Summarizing improvement cases in a resume and taking on a challenge at the presentation might be one of those "memorial trees."

(3) Visualize the improvement main points on the Rolling paper and record it in the keynote.
- It would have been worthwhile to record the progress of the activities

little by little.

- It's very easy to look back and put together later.
- Did you get your superior's comments?

(4) Don't forget to deploy it horizontally.

- When conducting horizontal deployment with related parties, ask your boss who is an adviser to cooperate if other departments are involved.

(5) Useful QC Methods / Analysis Tools

- Check sheets, photos, videos, etc.

[Procedure 2] **Check with each other for maintenance and improvement**

When the management method can be standardized in [Procedure 1], let's check with each other for maintenance and improvement and devise visual management so that it will not fade so that the effect can be passed on to the next generation.

(Procedure Points)

(1) Always keep in mind to make effective standards.

- Let's consider two types of quality management, result-based and factor-based.
- In particular, visual trend management using control charts and line graphs is effective.

(2) Useful QC Methods / Analysis Tools

- Line graphs, control charts, etc.

[Procedure 3] **Improvement announcements and remaining problems**

For the problem of chronic defects, which companies have suffered without a solution, the buried information in the field was muddy, and as a result of adding academic judgment as a statistical method, the root cause of the occurrence was found. Success stories with even more effective measures will strike the hearts of many. When you feel the atmosphere of the place, you will feel a sense of accomplishment and more confidence in the improvement activities that you have announced.

Next, we will clarify the remaining problems and aim for further improvement.

Procedure Points

(1) Look back on your activities.

- Thank you for your hard work. Let's discuss the four points of this activity with all the people concerned "to praise three points and make point one better."
- Everyone who ran desperately must have had a memorable point.

(2) Let's make a presentation of the activity report.

- Summarizing improvement cases in a resume and taking on a challenge at the presentation might be one of those "memorial trees."

[Improvement Example 11] Product A Dent defect Improvement-11
[Standardization of Measures Against Defective Dent]

The following three items have been standardized for improving Dent this time.

1. Method of dealing with material conditions and weaving of tool materials into mold design standards
2. Standardization of mold correction method
3. Incorporation into the mold corrector certification system

The above 1. The standardization of the internal structure and material improvement of the mold was carried out in Step 4, and the clues for countermeasures were found by the investigation of the phenomenon factor from the system and the detailed factor investigation from the cause system in Step 5.

On the other hand, regarding the above items 2. and 3., we analyzed the degree of influence on Dent in Step 8 by experimenting on the factors that we want to clarify by paying attention to the difficulty of mold adjustment mentioned in the factor arrangement in Step 6.

Both standardizations seem to have become important memorial trees for the members to maintain quality improvement based on the facts discovered this time and pass them on to the next generation.

[Another Standardization Left in the Factory]

For Kitaike andNishikawa, it seems that they were able to gain more satisfaction than achieving the goal in improving the defect of Dent, which was the most important issue of the factory. It was because the leader, Minamiyama, followed SOUTH-FLOW and introduced it to the members this time, that they became confident that they could tackle other chronic defects.

What was particularly impressive this time was that the people involved in improving the Dent defect were gathered by SOUTH-FLOW, and the employees who worked together for the first time were able to cooperate with each other.

[Dissolution of the Project Team]

The quality improvement project formed by the leader, Minamiyama, Chief Kitaike, and Nishikawa has succeeded in dramatically reducing the defective rate of new products, which has been a major problem since the start of mass production. They demonstrated outstanding teamwork in a period of about two and a half months.

They reported the results of the project to the director of Tokai Headquarters who directed the establishment of the project this time, and decided to dissolve the project, but it seems that each of the participating members has achieved great growth.

On the final day of the project, Mr. Kitaike was heading to the computer that he hated so much at the office.

Nishikawa: Kitaike, what on earth are you doing?
Kitaike: You can see what I'm doing. It's a computer.
Nishikawa: I understand that, but you hated computers so much...

Kitaike: That's right. However, when I saw Nishikawa on this case, I thought that I shouldn't have thrown away the computer nerd. When I get back to work, I want to tell the people in the field about this Southern thing.

Nishikawa: Well, it makes me happy when you say things like that.

[Improvement Software with SOUTH-FLOW]

Kitaike: By the way, Nishikawa, what exactly is this Rolling paper?

Nishikawa: As expected, as you woke up to using computers, you noticed a good point. This is a very convenient function, just click the Rolling paper button on the main menu of Figure 2.45.

Kitaike: Oh, something strange has started up.

Nishikawa: This Figure 2.46 is a Rolling paper. You are familiar with it, no? It's easy to organize the data because you can save the figures and tables created in the activity as images and list them later.

Kitaike: Isn't it the graphs and photographs we have examined so far? Somehow, it's like a commemorative photo album.

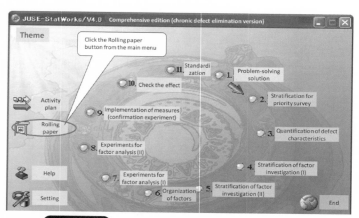

Figure 2.45 Main Panel of Improvement Software

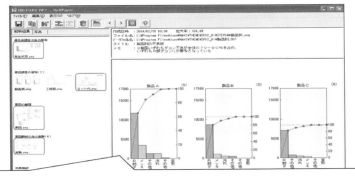

The charts saved as Rolling paper are listed here step by step, and the selected one is enlarged to the right. It can also be copied to the clipboard, so it can be used in other applications.

Figure 2.46 Rolling Paper for Improvement Software

[SOUTH-FLOW and Familiar Quality Improvement]

Nishikawa: "A commemorative photo album" is a good expression. By the way, the contents of the Rolling paper can be copied to Excel, Power-Point, etc., so it is also useful when creating reports and presentation materials such as Figure 2.47.

Kitaike: Hey, this isn't a QC circle resume, wait a minute, something is different from usual. Ah, what's this thing, each item is Southern something, isn't it? Moreover, even hints are included. If you paste the figure you checked up to the point where improvement has progressed, you can see the progress and there is a hint to check next, so it seems that it can also be used for circles.

Nishikawa: Kitaike, it's too early to be surprised. There are still some useful features hidden in this software. If you click on the step name here..., you can see Figure 2.48.

Kitaike: Oh, what's this little window?

Nishikawa: Here you can see all the steps of SOUTH-FLOW and the proce-dure. The procedure that has been completed will be displayed with a

SOUTH-FLOW How to Promote Chronic Defect Elimination Activities by SOUTH-FLOW

Chapter 2

check mark instead of the color of the icon, so you can see the progress level of the activity at a glance.

Kitaike: I see, if you look at this list, even if there are any omissions in the survey, it's obvious at a glance.

Nishikawa: Well, even if the activity is prolonged, especially in the case of a difficult problem, just have this, the review of the activity is perfect. It's like a "review help list."

Kitaike: Don't get carried away, silly. From now on, should I look at your progress of improvement with the chief of staff?

Nishikawa: Gee, I taught you something strange.

Kitaike: How stupid are you? I'm grateful that I'm going to peek in on you and give advice.

Nishikawa: Oh, is that so? Then I'm lucky.

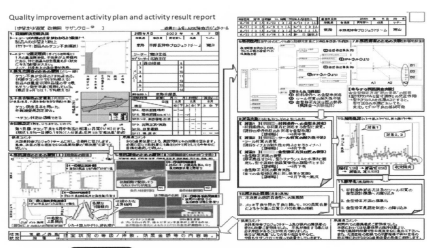

Figure 2.47 Report of Improvement Software (Keynote)

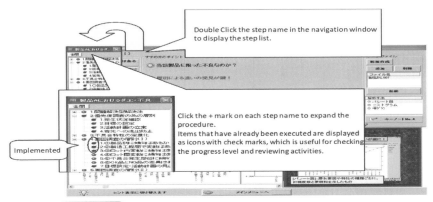

Figure 2.48 Progress Review List of Improvement Software

[Versatility of SOUTH-FLOW]

Minamiyama: Oh, Nishikawa and Kitaike will give you computer guidance. By the way, as you both know, this quality improvement project will be dissolved as of today. It seemed like a long time, and two and a half months happened in the blink of an eye, thank you very much. Thanks to you, I learned a lot again.

Nishikawa: This is where I learned a lot. Also, thanks to participating in this project, I was able to learn about "SOUTH-FLOW," so when I returned to the section, I wanted to use it for further quality improvement. Right, Kitaike?

Kitaike: Oh, I learned enough just to know the scenario of the Southern thing. I think this is also because the defective Dent has been improved. By all means, I'm embarrassed to say this is a memorial tree to my colleagues at other sites, but I'm thinking of telling them this cohesive improvement file. Well, but what is a computer? If there is a checklist that allows you to see the improvement history at a glance and check the progress of improvement with just one button, it is like a kind of navigation with Poka-yoke of improvement activities. I don't like the computer itself, but I wonder if it's okay to use it.

Minamiyama: As the improvement promotion secretariat of the headquarters, it would be nice if the improvement activities went well, and if the scenario was liked by operators and the future improvement became lively. Well then, let's go to the drinking party tonight!

Kitaike/Nishikawa: Oh ~.

Coffee Break

(6) Running Into an Old Classmate

▶**Question:**

Can you imagine a change in consciousness in your brain when you happen to meet an elementary school classmate and quality improvement while you are thinking about dinner on your way home from work?

▶**Example answer:**

First of all, we will let you know how the subconscious and manifesting consciousness explained below are related to memory and ideas.

(1) Awareness when you are alone in the crowd

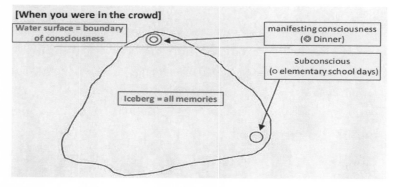

Dinner appears in the manifestation of consciousness. It's completely hidden in your mind, such as when you were in elementary school.

When trying to find out the true cause of a defect in improving chronic defects, the possibility of various discoveries is low if it is stored in the depths of the mind as in the above elementary school days.

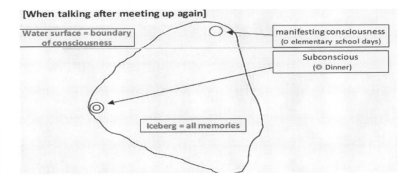

[When talking after meeting up again]

(2) Awareness when reuniting with an elementary school classmate and talking

When you talk to classmates you were close to after running into them, you naturally remember elementary school excursions and other classmates one after another. In addition, it may develop into planning a reunion that was unexpected before meeting up. On the contrary, if you don't meet each other, the story of the project will fade.

Even in quality improvement, when you discover an event of interest, many ideas and images come to mind. Also, as you discover several events, they are associated with each other and remembered by your brain, and the chances of creating ideas expand by accident.

(3) Awareness when meeting frequently after reuniting

If you continue to meet the classmate even once a week after meeting up, as shown in the figure, the elementary school days will be stored in a position that you can easily remember.

The more difficult it is to improve quality, the longer it takes to find the root cause. Also to more effectively relate some of the events gained during the improvement period. The shortcut to a solution is to work on improvement in a positive spirit with an awareness of problems for improvement through interest, repetition, and chain memory.

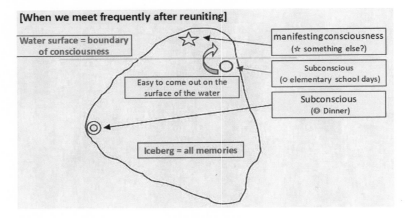

When eliminating chronic defects, by meeting with a close classmate (interest) even once a week (repetition), you can remind yourself of elementary school days one after another (chain memory). We would like to keep this in our brains so that it can be used for improvement.

Chapter 3 Useful Techniques for SOUTH-FLOW

"65% of quality problems in the market are recurring defects and it is difficult to eliminate chronic defects. Currently we try not to issue defective products to customers in the final inspection but the yield rate remains low. In order to reduce the defect rate, the most efficient and economically effective way is to improve their quality through the manufacturing process. However, the improvement method isn't circulating among the workers." I heard a complaint like this when we visited a major electrical equipment manufacturer.

In order to solve the problem of defects that occur chronically in the field, it is necessary to thoroughly investigate the phenomenon from all angles and grasp the true cause. To do so, not only a clear declaration of intent in management from the executives is a prerequisite, but also the presence of improvement leaders and project teams to eliminate chronic defects is essential. Also, both are necessary to build effective "problem-solving scenarios" for the company and to properly learn how to utilize the analysis methods that are weapons for implementing improvement.

In Chapter 3, we will discuss (1) Gauge R & R (GR & R), (2) Weibull analysis, (3) orthogonal array and factor layout experiment (exploratory design of experiments) as characteristic methods in SOUTH-FLOW. Figure 3.1 shows the outline and features of these three methods for utilizing SOUTH-FLOW. In the explanation of each method, we will focus on the necessary contents in the order of (1) method, (2) utilization method in SOUTH-FLOW and (3) analysis examples.

For more detailed explanations such as mathematical methods and formulas for calculating indicators and other QC methods, refer to the improvement software for operation help or the easy-to-understand SQC technical books below.

Please refer to Section 1.4 for the main QC methods used in SOUTH-FLOW. The output screen figures of this manual are in JUSE-StatWorks ® / V4.0 (chronic defect elimination version) which packages the concept of SOUTH-FLOW.

Method	Overview and Purpose	Result and Action	Related Clauses
Gauge R & R Analysis	When measuring multiple products, there is evaluation of the variation caused by multiple measurers, devices, methods and so on. It is possible to quantitatively judge whether or not the measurement variability (repeatability and reproducibility) is at the level of the standard.	If it is confirmed that there is no problem with the accuracy and reliability of the measurement system, one proceeds to the next step. If there is a problem with the repeatability of the measuring device or method, the next step cannot be taken unless the problem is removed. If it is a problem of human reproducibility, one limits the measurers and proceeds to the next step.	Chapter 2 Step 3 [Procedure 3] (p49)
Weibull Analysis	It is possible to estimate the state of failure occurrence such as initial and wear failure of parts, and Mean Time To Failure (MTTF) and statistically determine the cause of failure by plotting the data on the Weibull probability paper.	The cause is inferred by comparing differences in the tendency of defects that occur due to the difference in the life of molds and jigs.	Chapter 2 Step 5 [Procedure 5] (p78)
Exploratory Design of Experiments	It is possible to efficiently search for factors that affect the characteristics from a plurality of "possible factors" and find the optimum conditions. In particular, when there are many factors, the orthogonal array and the factor placement experiment are often used in combination. If there are few factors, a factor placement experiment including a confirmation experiment is sufficient.	One efficiently narrows down "possible factors" for quality characteristics to be improved and confirms "factors that contribute" to the results with the degree of influence. Next, the optimum conditions are quantitatively searched from among the "contribution factors".	Chapter 2 Step 7 [Procedure 1] (p97) Step 8 [Procedure 1] (p102)

Figure 3.1 List of "Useful Techniques" for SOUTH-FLOW

3.1 Gauge R & R Analysis
- Checking the Reliability of Measurement Data -
(1) What is Gauge R & R (GR & R)?

Gauge R & R analysis is a measurement system (a series of processes for obtaining measured values including measuring equipment, measurer and operations) for data repeatedly measured by multiple evaluators for multiple parts (products and parts). It is a method to evaluate the variation (including the method, etc.) and quantitatively judge whether the measured value is a stable collection.

For example, it is important to keep track daily of whether the control items required for process design, process control and product control are being measured correctly.

Gauge R & R is used in the field to analyze how much the measurement capacity is or what needs to be improved at the work site. In other words, it is used to evaluate the reliability of the measurement system or the measurement value itself before grasping the current situation.

In the MSA (Measurement System Analysis) related manual of QS-9000 established by the American automobile industry, Gauge R & R has been adopted as a quality requirement for suppliers of parts, and has been widely used in Japan as of late.

Gauge R & R in SOUTH-FLOW is recommended for use when confirming the reliability of quality characteristic evaluation as [Procedure 3] "Implementation of Gauge R & R (GR & R)" in Step 3.

(2) How to use it in SOUTH-FLOW

When multiple products are repeatedly measured by multiple evaluators, the obtained data may vary depending on the measuring device and method (repeatability), and variations due to different evaluators (reproducibility). SOUTH-FLOW uses Gauge R & R as a method of statistically determining whether measurement reliability is sufficient for the quality characteristic standards.

Chapter
3

Useful Techniques for SOUTH-FLOW

137

SOUTH-FLOW judges the reliability of measurement by comparing the Gauge R & R (%) with the size of the standard width. The criteria is as the following.

If the difference between Gauge R & R and the standard width is:

- less than 10%, it is considered a pass.
- 10% to 30%, it is considered a conditional pass.
- 30% or more, it is considered a fail.

If this number is more than 30% and fails, this measurement cannot accurately represent the inferiority of the product. Therefore, it means the data obtained after Step 4 will also be unreliable. If the reliability of the measurement fails, it is necessary to review the measurement method.

(3) Analysis Example

At JUSE Machinery Works, a full inspection off-line is carried out as a shipping inspection due to variations in the dimensions of the products inspected in-line. In order to eliminate the wastefulness of doing a full inspection twice, they decided to investigate the current situation. Therefore, they conducted a Gauge R & R to confirm whether the measured values are reliable. In-line workers measure multiple samples twice. There are eight products and four evaluators, therefore the total number of data is 64 (Figure 3.2).

Notice

When collecting data, make sure that the measurer does not know the previous day's data. Also, be careful not to damage the sample prepared in Gauge R & R, such as deformation, during measurement.

If there are multiple measuring machines, prepare multiple samples near the median or the upper and lower limits of the standard instead of the sample used in the Gauge R & R when checking the difference between the measuring machines.

138

	● S 1	● C 2	● C 3	● N 4
	Sample Name	Part	Evaluator	Measured Value
● 1	1	P1	Evaluator A	5.70
● 2	2	P1	Evaluator A	5.40
● 3	3	P1	Evaluator C	5.50
● 4	4	P1	Evaluator C	5.30
● 5	5	P1	Evaluator D	4.90
● 6	6	P1	Evaluator D	5.20
● 7	7	P1	Evaluator B	4.70
● 8	8	P1	Evaluator B	4.80
● 9	9	P2	Evaluator A	9.80
● 10	1 0	P2	Evaluator A	9.50
● 11	1 1	P2	Evaluator C	8.40
● 12	1 2	P2	Evaluator C	9.00
● 13	1 3	P2	Evaluator D	8.80
● 14	1 4	P2	Evaluator D	9.10

Figure 3.2 **Example of Data Table (One Section)**

1. Enter the part name, measurer and measured value. Measurements are performed multiple times under the same conditions.
2. Select Gauge R & R analysis.
3. Select the method based on the mean and range as the method for estimating the dispersion component.
4. Perform ANOVA (analysis of variance).
- Analyze Gauge R & R by ANOVA table.
- Enter tolerances or upper and lower limit standard values (Figure 3.4).

Figure 3.3 **Analysis Procedure for Gauge R & R Analysis**

The figures in the Gauge R & R table in Figure 3.5 are stratified by item, as in the graph in Figure 3.6, so you can visualize where the problem lies.

《**What Was Learned**》

From Figure 3.5, the Gauge R & R row tolerance ratio (%) is 63.7% which

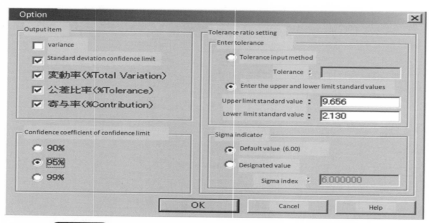

Figure 3.4 Setting Standard Values, Confidence Intervals, etc.

ndc: 2 (2.957) Sigma index: 6.00 Tolerance: 7.526

ndc : 3 (3.007)	Sigma index: 6.00	Tolerance : 7.526				
Dispersed Component	Standard Deviation	90% Confidence Limit		Total variation(%)	Tolerance (%)	Contribution (%)
		Lower Limit	Upper Limit			
1 Repeatability	0.230	0.192	0.291	12.1	18.3	1.5
2 Reproducibility	0.714	0.395	2.126	37.6	56.9	14.1
3 Parts Evaluator	0.295	0.183	0.448	15.6	23.5	2.4
4 Gauge R & R	0.806	0.572	2.142	42.5	64.3	18.0
5 Parts	1.719	1.082	3.198	90.5	137.1	82.0
6 Total	1.899			100.0	151.4	100.0

Figure 3.5 Gauge R & R table

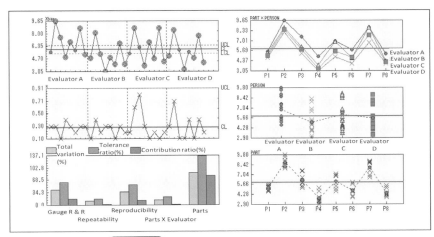

Figure 3.6 Graph Display of Gauge R & R

means it is unsuccessful when used for a full inspection in-line. In addition, since the reproducibility tolerance ratio (%) is 60.6%, it was found that the reliability of measurement can be improved by reselecting the measurement operator. In this analysis case, you can find suggestions for improvement by focusing on the differences between the other measurement tasks with respect to the similarities between the measurement tasks of Evaluator C and Evaluator D from the graphs stratified and plotted by measurement operators in Figure 3.6.

It is also necessary to promote the standardization of measurement work here, but since it is necessary to confirm the reliability of measurement in Step 3 of SOUTH-FLOW aiming at source countermeasures, first of all, the selected measurer (C or D) should be asked. Request an investigation and proceed to Step 4.

Chapter
3

Useful Techniques for SOUTH-FLOW

3.2 Weibull Analysis
- Grasping and Quantifying the Failure Status of Parts -
(1) What is Weibull Analysis?

Probability papers based on a wide variety of distributions are used for reliability data analysis, but the most commonly used is the Weibull probability paper.

As shown in Figure 3.7, the Weibull probability paper plots data on a graph to estimate values such as assumed distribution parameters (shape parameter m and scale parameter η) and MTTF (Mean Time To Failure) of parts. It is possible to statistically consider the condition of failure occurrence.

The Weibull Probability Paper has the cumulative failure probability value F(t) on the vertical axis and the time to failure on the horizontal axis, plots the data and estimates the parameters from the fitted linear relationship (slope of a line, m). As shown in Figure 3.8, the instantaneous failure rate λ (t), which is an important measure in reliability technology, can classify the phenomena in which failures occur according to the value of m and is one guideline for investigating the causes of failures.

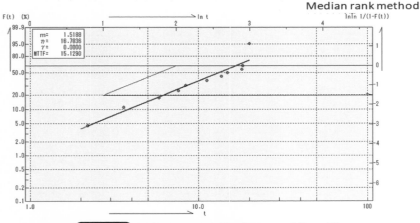

Figure 3.7 Weibull Probability Paper and Data Plot

When m <1 ...... Reduced fault rate type (initial fault)

When m = 1 .... Fixed fault rate type (random fault)

When m> 1 .... Increasing fault rate type (wear fault)

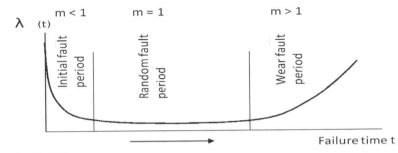

Figure 3.8 **Momentary Failure Rate and Trend of Occurrence Over Elapsed Time**

For example, it is possible to know the temporal trend of failure occurrence from the value of the instantaneous failure rate, m.

In this way, the value of the shape parameter m is estimated from the plot of the data on the Weibull probability paper, and it is classified from the initial failure (m <1), accidental failure (m = 1) and wear failure (m> 1). You can see what type of product failure is occurring or if it has reached the end of its life.

In SOUTH-FLOW, Weibull Analysis should be used at "[Procedure 3]" "Differences by Mold/Jig (Machine-2)" in Step 5 to confirm the difference in the trend of defects to occur depending on the life of the mold and tool.

(2) How to Use Weibull Analysis in SOUTH-FLOW

Weibull analysis is used to investigate the life of parts due to wear of molds and jigs. The main indicators used here are the shape parameter, (m), by which the causative system of the failure is known, MTTF (μ) as the mean time between failures and the standard deviation (σ).

As shown in Figure 3.8, for the shape parameter m, if the value of m is less than 1, which indicates 0.6, it means that the tools such as molds and jigs are eas-

143

ily chipped and the strength is insufficient from the beginning of use. Therefore, it leads to measures such as changing to a material that is resistant to chipping.

On the other hand, if the value of m is bigger than 1, like 2.0, it indicates that the tool will eventually wear out as it is used. Therefore, you should take measures such as changing to a material that is less likely to wear away.

MTTF (μ) and standard deviation (σ) as life expectancy determine which failure mode or component should be prioritized for improvement.

(3) Analysis Example

At Manufacturer A, there was an opinion from the work site that if tools were used for a long period of time, defects would easily occur. Therefore, this company decided to conduct a reliability test (accelerated test) for confirmation. In order to understand the failure status of 15 test pieces, they decided to conduct a test for 8200 minutes (137 hours) and analyze the data (Figure 3.9). What kind of knowledge did they get as a result of data analysis? There were 15 test pieces, five of which did not fail and terminated normally. For the failed data, the failure time was measured and the cause of the failure was identified. There are two types of failure causes, A and B, which are classified and described according to failure mode.

Figure 3.10 shows the main steps in performing the Weibull analysis.

《What Was Learned》

From the analysis results, the shape parameter m of the cause of failure A was 0.851, which was almost an accidental failure, but the value of m of the cause of failure B was as large as 3.21, indicating the pattern of wear failure. Therefore, especially for failure cause B, the number of failures is likely to increase rapidly as the test time increases and it is necessary to take measures such as changing to a wear-resistant material or replacing parts regularly. It is probable that according to the opinions from the site there was an awareness of the occurrence of defects due to failure cause B.

● S 1 Sample name	● N2 Time to failure	● C 3 Failure mode
1	660	A
2	1500	A
3	2700	A
4	3400	B
5	3800	A
6	4200	Normal termination
7	4400	B
8	4700	B
9	5200	Normal termination
10	5600	B
11	6600	B
12	7200	Normal termination
13	7200	Normal termination
14	7600	B
15	8200	Normal termination

Figure 3.9 Data Table (Time of Failure, Failure Mode)

(1) Enter information for the time of failure, termination status, and failure mode for each sample.
(2) Specify data and make initial settings (Figure 3.11).
(3) Display Weibull probability paper (Figure 3.12).
(4) For multiple items compare m, MTTF (μ), σ, etc. (Figure 3.13).

Figure 3.10 Analysis Procedure for Weibull Analysis

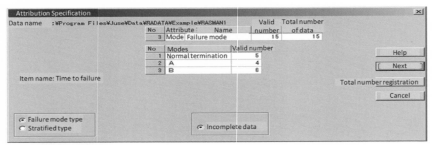

Figure 3.11 Initial Settings for Weibull Analysis (Stratification Types, Number of Products Manufactured, etc.)

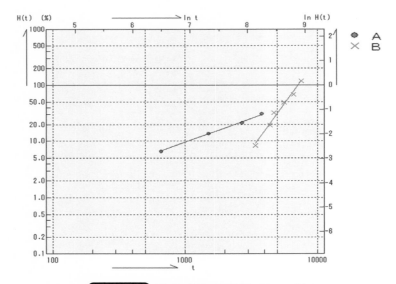

Figure 3.12 Weibull Probability Paper Plot

Failure cause	n	N	m	η	γ	MTT(B)F(μ)	σ	10 percent point
1 A	4	15	0.8561	15537.2255	0.000	16829.1558	19734.4365	1121.4255
2 B	6	15	3.2122	7168.6177	0.000	6421.7863	2195.1735	3557.8085

Figure 3.13 List of Obtained Parameters

3.3 Orthogonal Array and Factor Placement Experiment (Exploratory Design of Experiments)

- Efficient Factor Narrowing and Searching for Optimal Conditions -

(1) What is the orthogonal method and the factor layout experiment?

First, the orthogonal array is a method that can be performed with a smaller number of experiments for a relatively large number of factors. This is done by taking advantage of the property of which all combinations of the levels appear the same number of times for multiple factors in the causative system. Although there are several types of orthogonal arrays, two-level orthogonal arrays such as L_8 and L_{16} and three-level orthogonal arrays such as L_9 and L_{27} are often used.

Experimental design originates from a methodology for experimental design devised by Fisher, an engineer at the Rothamsted Experimental Station in the United Kingdom around 1925. The purpose of experimental design is to efficiently discover optimal conditions and values by "changing the state of factors that are thought to affect the results, setting a certain level and comparing the results in order to achieve a certain purpose."

Here, in particular, an orthogonal array experiment (experiment to create a large net) to narrow down the true cause from multiple elements (factors) and a factor layout experiment to find the optimal level from the narrowed down factors are performed; it is described that effective exploratory experimental design can be performed by combining them.

First, the orthogonal array is a method that can be performed with a smaller number of experiments for a relatively large number of factors by taking advantage of the property of which all combinations of the levels appear the same number of times for multiple factors in the causative system. Although there are several types of orthogonal arrays, two-level orthogonal arrays such as L_8 and L_{16} and three-level orthogonal arrays such as L_9 and L_{27} are often used.

Considering man-hours, resources and work time, experimental design is effective for parameter design and the search for optimal conditions in short-term development. By combining the use of orthogonal array and factor layout

147

No	Source	Sum of squares	DF	Mean square	F	Test	P-value(upper)	Contribution rate
1	A	1.27125	3	0.42375	8.268	**	0.003	16.565
2	B	4.73250	2	2.36625	46.171	**	0.000	68.631
3	A x B	0.12750	6	0.02125	0.415		0.856	0.000
	Error e	0.61500	12	0.05125	-			14.805
	Total	6.74625	23	-	-			

Figure 3.14 Example of Two-Way ANOVA Table (Repeated Twice)

experiment, effective results can be obtained with less man-hours.

《ANOVA Table and Allocation》

ANOVA (analysis of variance) is a method of analyzing the variation of characteristic values by factor and finding out what factors (factors and interactions) have a particularly large effect on the error (usually the variance of each factor). The error variance is compared and tested by the magnitude of the F value to select the important factor. A table that summarizes the analysis results as shown in Figure 3.13 is called an analysis of variance table. For example, as shown in Figure 3.14, there are two * marks in the "Test" column of Factor A and Factor B and it can be determined with a 99% probability that this factor has an effect on quality characteristics. If there is only one * mark in the "Test" column, it can be concluded that there is a 95% chance of influence.

(2) Usage in SOUTH-FLOW

In SOUTH-FLOW, if the number of "possible factors" is up to six (four factors, two interactions, etc.), we recommend L8,which can be easily assigned using the basic composition table by eight experiments instead of $2^6 = 64$ experiments with six combinations.

Furthermore, if you want to find the optimal conditions after narrowing down the target factors to a few, we recommend performing a factor layout experiment. There are one-way (one factor), two-way (two factors) and three-way (three factors) ANOVAs depending on the number of factors but the most commonly used is the "two-way ANOVA."

In SOUTH-FLOW it is recommended to perform an orthogonal array experiment to search for "factors that contribute to characteristics" from "possible factors" in the experiment (1) for factor analysis in Step 7. It is also recommended to combine factor layout experiments to find the optimal conditions from among "contributing factors" in Step 8.

In order for the experimental design to be successful, pay attention to the following items as well.

1) Before conducting the experiment, narrow down the extraction of controllable factor candidates, interactions, etc. as much as possible.

2) If the degree of influence cannot be confirmed, pool factors and interactions as necessary. If the variance ratio F_0 value is 2.0 or less compared to the variation in the error term, the corresponding factor is pooled from the interaction.

3) Examine whether the experimental results meet the target value. Examine both the next level combination and the optimal conditions.

4) Investigate the final results by the confirmation experiments with unique technology.

Since the L_8 orthogonal array requires less experimental data, the number of factors can be fewer and the detection power for testing significance also less, so the factors and interactions at the survey stage can be narrowed down and experiments can be performed repeatedly. Orthogonal experiments are used to investigate the presence or absence of influential factors and interactions, followed by factor layout experiments to find optimal conditions. If the factors and interactions are not sufficiently narrowed down by prior investigation and analysis, select L_{16} or L_{32}, which has more experimental data than L_8. Also, in order to know the peak, select experiments such as L_9 and L_{27} of the three-level system.

(3) Analysis Example -1

Resin Manufacturer B has decided to improve adhesive strength. There are four factors related to adhesive strength: A, B, C, and D and the technically possible interactions were combinations of AB and AC. Therefore, since there are a total of six factors, four factors and two interactions, they decided to conduct an L8 orthogonal array experiment. The target was 7.0 or higher in the 95% reliability interval.

Exhibit 3.15 shows the main steps in performing an L8 orthogonal array experiment.

《What Was Learned》

- As a result of the experiment, B and the combination of AC were significant (ρ value 5% significant), and the optimal conditions were A2, B2, and C1.
- The population average of the tensile strength under this optimal condition was 9.5 and the 95% confidence interval of the population average was 7.09 to 11.91, clearing the target value of 7.0 in the 95% confidence interval.

(1) Assign factors and interaction terms to each column while looking at the component symbols (Figure 3.16).

(The experiment number is "No." in the rows, Column 1 is Factor A, Column 2 is Factor B, Column 3 is AB, Column 4 is Factor C, Column 5 is AC, Column 7 is Factor D, Column 6 is the error e.)

(2) Randomize so that the trend due to the experimental order does not appear.

(3) Conduct experiments and collect data.

(4) Enter the experimental design table and data (Figure 3.17).

(5) Display the analysis of variance table (Figure 3.18).

(6) Since there is no * in the test for any factor, pool (F_0 value <2) and display the analysis of the variance table again (Figure 3.19).

(The F value of interaction AB is 1.00 and 2 or less, which is almost the same as the magnitude of the error, so pooling)

(7) Select the factors marked with * in the analysis of the variance table and display the estimated values (Figure 3.20).

(Select Factors A, B, C, AC because there is a * in the B and AC lines)

(8) Furthermore, the factor effect plot diagram is displayed (Figure 3.21).

Figure 3.15 L8 Orthogonal Array Experiment Procedure

Number \ Column	1	2	3	4	5	6	7
1	1	1	1	1	1	1	1
2	1	1	1	2	2	2	2
3	1	2	2	1	1	2	2
4	1	2	2	2	2	1	1
5	2	1	2	1	2	2	2
6	2	1	2	2	1	1	1
7	2	2	1	1	2	1	1
8	2	2	1	2	1	2	2
Symbol	a	b	ab	c	ac	bc	abc
Assignment	A	B	A×B	C	A×C	e	D

Figure 3.16 L8 Orthogonal Array (Basic Components and e: error)

● S1 Experiment No.	● N2 Factor A	● N3 Factor B	● N4 Factor C	● N5 Factor D	● N6 Data
s1	1	1	1	1	2.0
s2	1	1	2	2	4.0
s3	1	2	1	2	5.0
s4	1	2	2	1	6.0
s5	2	1	1	2	7.0
s6	2	1	2	1	2.0
s7	2	2	1	1	9.0
s8	2	2	2	2	7.0

Figure 3.17 Design of Experiments and Data (Level Values and Data for Each Factor)

Source	Sum of squares	DF	Mean square	F	Test	P-value(upper)
A	8.000	1	8.000	16.000		0.156
B	18.000	1	18.000	36.000		0.105
C	2.000	1	2.000	4.000		0.295
D	2.000	1	2.000	4.000		0.295
AB	0.500	1	0.500	1.000		0.500
AC	12.500	1	12.500	25.000		0.126
Error	0.500	1	0.500			
Total	43.500	7				

Figure 3.18 ANOVA

Source	Sum of Squares	DF	Mean square	F	Test	P-value(upper)
A	8.000	1	8.000	16.000		0.057
B	18.000	1	18.000	36.000	*	0.027
C	2.000	1	2.000	4.000		0.184
D	2.000	1	2.000	4.000		0.184
AC	12.500	1	12.500	25.000	*	0.038
Error	1.000	2	0.500			
Total	43.500	7				

Figure 3.19 ANOVA (After Pooling AB)

Factor ABC	Population	Confidence interval		
		Lower (95%)	Upper (95%)	
111	2.0	-0.41	4.41	min
112	3.5	1.09	5.91	
121	5.0	2.59	7.41	
122	6.5	4.09	8.91	
211	6.5	4.09	8.91	
212	3.0	0.59	5.41	
221	9.5	7.09	11.91	max
222	6.0	3.59	8.41	

Figure 3.20 Estimates and Confidence Intervals

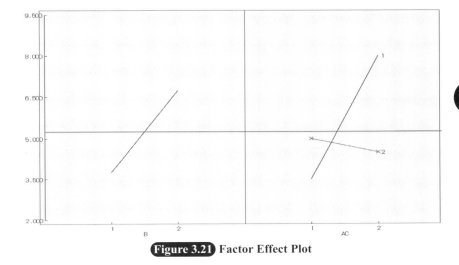

Figure 3.21 Factor Effect Plot

(4) Analysis Example-2

Chemical manufacturer Q manufactures chemical Z. Yields have fall-
en short of the target of 80%, putting pressure on profits. Therefore,
in consideration of past statistical knowledge and unique technology,
factor layout experiments are conducted for Factor A (type of principal
component) and Factor B (amount of additive) to determine the optimal conditions.

Figure 3.22 shows the data obtained in the binary factor layout experiment.
A is two levels, B is five levels, and it is repeated twice. In addition, Figure 3.23
shows a factor-effect plot, in which A, B, and AB are subtly intertwined and affect the yield.

《**What Was Learned**》

In the combination of this experiment, the optimal condition for maximiz-
ing the yield is A2B2 (A2 = main component 2, B2 = addition amount 30.0 g),

Factor	B1 (27.5g)	B2 (30.0g)	B3 (32.5g)	B4 (35.0g)	B5 (37.5g)
A1 (main component 1)	73.095	78.024	80.238	75.922	80.272
A1 (main component 1)	74.020	74.315	80.746	82.706	81.506
A2 (main component 2)	81.224	84.596	78.791	71.546	61.314
A2 (main component 2)	77.984	84.545	80.888	70.318	62.873

Figure 3.22 Design of Experiments with Two Elements and Data

and the point estimation of the population average of the yield at that time is 84.571%. This confidence interval increased from 81.395 to 87.746 (clearing the target value of 80%), and there was an interaction between Factor A and Factor B, and from the BA estimated value plot, A1 (main component 1). Then, it was found that the yield peaked around an addition of B = 37.5 g, and at A2 (main component 2) around an addition of B = 300 g.

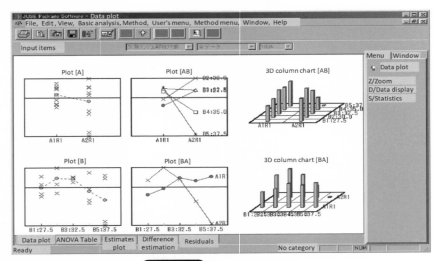

Figure 3.23 Factor Effect Plot

Chapter 4 Success Stories Using SOUTH-FLOW

Chapter 4 introduces successful cases of eliminating chronic defects using SOUTH-FLOW. The following examples were created based on actual activities of three companies in different industries.

4.1 Improvement of Weighing Defect Rate of Subdivided Packaging at Chemical Manufacturing Company N

4.2 Improvement of Defective Assembly Dimensions of Board Parts in IT-Related Company C

4.3 Development of New F Substrate Adhesive X at Resin Manufacturing Company A

4.4 What Was Learned From the Success Stories in These Three Different Industries

In the case of 4.1, we worked on improvement with commonplace repair work of many years as the theme. In particular, according to Step 3: "Survey on Measurement Reliability" by GR & R and on-site confirmation of key people based on SOUTH-FLOW, we found a factor in the difference in the adjustment value of manufacturing conditions and linked it to improvement.

In the case of 4.2, we worked on quality improvement by providing support and guidance from Japan regarding quality defects at a Chinese factory. Beginning with the selection of defects to be addressed by SOUTH-FLOW's Step 2: "Priority Survey," it ends with the successful improvement of their production quality by discovering the factors of manufacturing problems of local suppliers in China particularly with Step 5: "Investigation of Differences by Line."

In the case of 4.3, we worked on development by applying SOUTH-FLOW to the unresolved development products until now. Factors were selected by Step 4: "Investigation of Similar Products" in SOUTH-FLOW, and optimal conditions were discovered by the "orthogonal array experiment" in Step 8 and linked to improvement.

In 4.4, we have summarized the commonalities and differences in the two major SOUTH-FLOW processes (the process of searching for the root cause and the process of forming the strongest team) for the success stories in the three different industries introduced in this chapter. We will introduce the features of the three cases, including the main QC methods used for improvement

4.1 Improvement of Weighing Defect Rate of Subdivided Packaging at Chemical Manufacturing Company N

The improvement leader is Mr. T from the Facilities Division of the Manufacturing Department. At Company N we used SOUTH-FLOW to work on reducing the measurement defect rate which has been commonplace for many years.

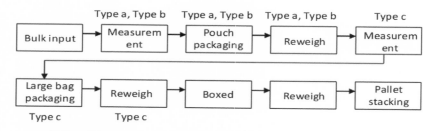

Figure 4.1 Schematic Diagram of the Product Packaging Process

First, the bulk of the material is put into the hopper, transferred to the weighing machine by the bucket conveyor, weighed, put into the packaging machine and packaged in sachets. After that, it is inspected by a weight checker and then packed in a large bag. In large bag packaging, as with small bag packaging, small bags are transferred to a weighing machine by a packet conveyor to be weighed, put into a packaging machine, packed in a large bag and then boxed and palletized via a weight checker.

> ## Step 1: Problem-Solving Solution
> ## [Scenario selection for pursuing the root cause]

[Procedure 1] **Solution Theme Selection**

At Company N, a large number of excess and lightweight defective packaging products occur during subdivided packaging.

Defective products are unpacked and repacked but there is much loss in man-hours and materials and naturally for many years they have given up on repair. However, his superior ordered Mr. T to improve this as a leader because repair was wasteful. When the people involved in the manufacturing of subdivided packaging were asked, they said that defects in subdivided packaging were largely due to the person in charge at the site and standardization would be necessary to maintain stable production. This theme was chosen because it was thought that a clue for improvement could be obtained by visualizing on-site information using SOUTH-FLOW.

[Procedure 2] **Registration of Theme Name**

The theme name is to reduce the weighing defect rate of subdivided packaging.

> ## Step 2: Stratification for Priority Survey
> ## [What to Start With]

[Procedure 2] **Setting Goals**

As a goal we set C_{pk} to 1.33 or higher in order to aim for an order of magnitude quality (weighing defect rate 1/10). The in-house standard for subdivided packaging is 30 to 32 g.

[Procedure 3] **Preventing the Outflow of Defective Products to Customers**

To prevent outflow to customers, we reweigh all of them off-line.

Step 3: Quantification of Defect Characteristics
[Reliability Confirmation of Quality Characteristic Evaluation]

[Procedure 3] **Implementation of Gauge R & R (GR & R)**

We conducted GR & R on the reliability of the weight checker that checks the weight of the product. First of all, we weighed 30 samples of 30g and 300g each with a standard scale and made them. Next, the conditions (air conditioning ON/OFF) were changed using 30 g for Type a and Type b and 300 g for the Type c, which are all subdivided packages, and each measurement was performed once on three people per day for three days, measuring the samples with a machine.

Here, the process of checking GR & R by dividing the air conditioning into ON and OFF will be explained.

The first day of measurement started with sample measurement from the Type a measuring machine, but when a worker turned off the air conditioning during measurement, it became apparent that the measured value of the same 30g pouch was different from when it was turned on. Therefore, it was decided to collect both air conditioning ON and OFF data for all measuring instruments.

Figure 4.2 shows a reliability comparison for each measurement system that summarizes the GR & R results. From Figure 4.2, it was found that air conditioning is clearly a factor that deteriorates the reliability of measurement in the Type a and Type b measuring machines used for subdivided packaging. On the other hand, it was also found that the Type c measuring machine used for the large bags was not affected by air conditioning.

	Type a		Type b		Type c	
Air conditioning	ON	OFF	ON	OFF	ON	OFF
GR&R(%)	49.58	12.42	95.07	43.79	20.11	21.47

Figure 4.2 Comparison of Measurement Reliability for Each Measurement System

Comparing the GR & R values with the air conditioning turned off there, as can be seen from Figure 4.2, it increased in the order of Type a (12.4%) < Type c (21.5%) < Type b (43. 8%). It was clear that some factor, due to the difference in the measurement system, had some effect. As in SOUTH-FLOW, if the measurement is unreliable, it is necessary to review the measurement, so it became impossible to proceed with activities to reduce the measurement defect rate of subdivided packaging.

Therefore, we asked the people involved in the manufacturing department, including the superior, to gather and explain what was learned from the on-site measurement reliability survey. The contents explained were as follows.

(1) From the schematic diagram of the product packaging process, there are two measuring instruments for Type a and Type b subdivided packaging and one for Type c large bag packaging.

(2) Due to the difference in the air conditioning being on and off, the GR & R values of the two machines Type a and Type b for subdivided packaging should be respectively improved to one-fourth and one-half.

(3) One Type c machine for large bags should not be affected by the difference in air conditioning being on and off.

(4) When the air conditioning is off, the GR & R value increases in the order of Type a (12.4%) < Type c (21.5%) < Type b (43.8%).

(5) According to SOUTH-FLOW, improvement of weighing defects in subdivided packaging should not be promoted in such an unreliable state.

Those who assembled that are familiar with the site looked overhead and noticed the following regarding air conditioning.

i) A windshield is installed on the air conditioner above the Type c measuring device and wind does not blow directly.

→ It is unknown why it was only installed here because no one remembers what happened.

ii) There is no windshield on the air conditioners above Type a and Type b, and wind blows directly; Type a in particular is directly under the air conditioner and is clearly closer to the air conditioner than the Type b.

→ Regarding the windshield near the Type a and Type b machines, the installation was decided on the spot.

iii) The basic structure of the measuring machine is the same for Types a, b and c but the operator is supposed to adjust the condition setting values around the measuring machine and although the values are set within the allowable adjustment range, all settings were different.

At this meeting, because turning off the equipment air conditioning will affect the quality of the product, it was decided to investigate after first installing windshields for the air conditioner.

Step 4: Stratification for Factor Investigation (I)
[Phenomenon] [Finding factors from the result system]

[Procedure 4] **Is There a Tendency for Lot-to-Lot Fluctuations?**

After installing the windshield for the air conditioner, the confirmation in Step 4 was carried out. Figure 4.3 shows the X-R control chart to indicate the trend between lots of the air conditioning measurement system of Type a.

No particular trend was seen.

Figure 4.4 shows the X-R control chart to indicate the trend between lots of the air conditioning measurement system of Type b.

No particular trend was seen.

Step 5: Stratification for Factor Investigation (II) [Details]
[Search for factors from the causative system]

[Procedure 4] **Difference by Line (Machine 1)**

Figure 4.5 shows a histogram comparing the process capabilities of Type a and Type b. From Figure 4.5 it was found that the process capability of Type a was $C_{pk} = 2.01$ and cleared the target but that Type b was slightly short of the target with a value of $C_{pk} = 1.06$.

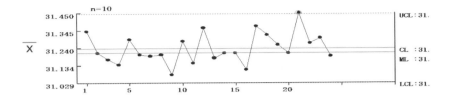

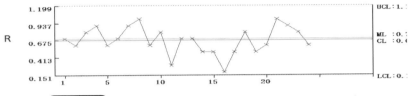

Figure 4.3 Measurement System Type a Time Series Confirmation X-R Control Chart

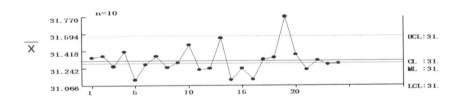

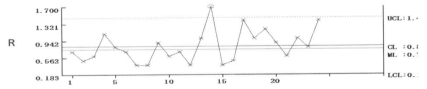

Figure 4.4 Measurement System Type b Time Series Confirmation X-R Control Chart

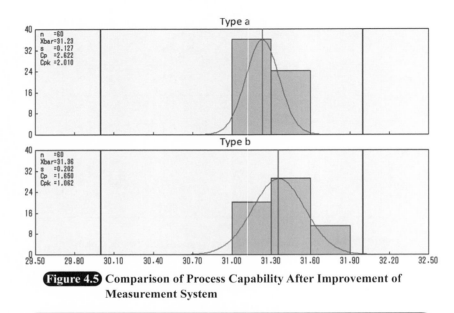

Figure 4.5 Comparison of Process Capability After Improvement of Measurement System

Step 6: Sorting Out the Factors
[Organize factors with the strongest members using facts as hints]

[Procedure 1] Arrangement of the Relationship Between Defect Characteristics and Factors

In Step 6, key people were asked to gather and organize the factors for measurement defects into an Ishikawa diagram based on factors, intuition, and experience up to Step 5 as shown in Figure 4.6. As expected, many factors related to the weighing machine were mentioned but the workers also mentioned the level of skill related to the tricks and tips for adjusting the conditions.

Steps 7 and 8: Experiments for Factor Analysis (I), (II)

The factors this time were that the reliability of the measurement was low due to the lack of a windshield on the air conditioner, and that the adjustment of the conditions around the Type b weighing machine was not optimized. So the

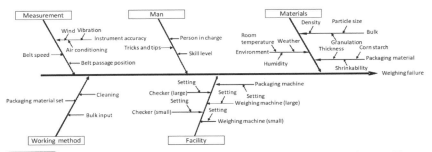

Figure 4.6 Organization of Factors for Improper Measurement Ishikawa diagram (excerpt)

experiment was omitted having been judged as unnecessary.

Step 9: Implementation of Measures
[Introduction of optimal conditions to mass production]

[Procedure 2] Implementation of Mass Production Measures

The measures incorporated into mass production were the installation of windshields on the air conditioners near the Type a and Type b measuring machines and the standardization of the conditions around the Type b weighing machines according to the conditions around Type a (details of work standards and conditions omitted).

Step 10: Effect Confirmation
[Confirmation of effects after introduction of mass production measures]

[Procedure 1] Confirmation of the Effect of Improvement

Figure 4.7 shows a histogram for comparing process capability before and after implementing measures around the Type b weighing machine. The process capability, which had a C_{pk} of 1.06 before improvement became 1.88 after improvement, and the target could be cleared.

Chapter 4

Success Stories Using SOUTH-FLOW

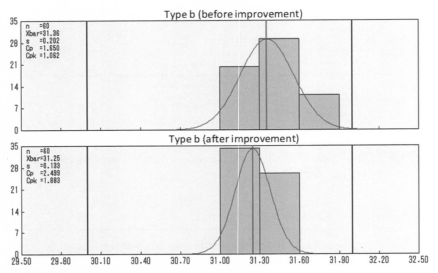

Figure 4.7 Histogram of Comparison Before and After Improvement Around Type b Weighing Machines

Step 11: Standardization
[Succession of improvement measures to the next generation]

Omitted from print.

～ Impressions About Using SOUTH-FLOW ～

The manufacturer of the weighing machine insists that the measuring instrument has some errors since it is only a tool for checking, so no special measures have been taken so far.

This time, SOUTH-FLOW's Step 3: "Confirmation of Reliability of Quality Characteristic Evaluation" was a good opportunity to review the weighing machine.

The GR & R method that we used for the first time revealed at an early state that the weighing machine was not functional for the standard. Therefore, as a result of seriously considering measures for weighing machines without making detours, we were able to ensure the reliability of weighing. I would like to continue improvement using SOUTH-FLOW.

(interview with Mr. T, Improvement Leader of Company N)

Chapter 4

Success Stories Using SOUTH-FLOW

4.2 Improvement of Defective Assembly Dimensions of Board Parts in IT-Related Company C

The improvement leader is Mr. S from the Quality Assurance Division. Company C provides support and guidance from Japan regarding quality defects at its Chinese factory. At that time the defective rate of product B was high and affecting profits. In order to reduce it, SOUTH-FLOW was used.

The manufacturing process is shown in Figure 4.8.

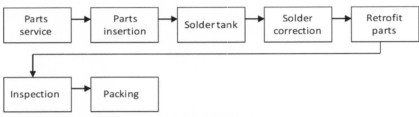

Figure 4.8 Schematic Diagram of the Printed Circuit Board Assembly Process

Step 1: Problem-Solving Solution
[Scenario selections for pursuing the root cause]

[Procedure 1] Solution Theme Selection

The reason for selecting the theme is that when starting local production in China, the defective rate of product B is high and there is a large gap with the income and expenditure plan. It was decided that it is better to carry out quality improvement together with the Chinese support members who are carrying out start-up support locally.

[Procedure 2] Registration of Theme Name

The theme name was the extraordinary quality improvement for printed circuit board product B.

Step 2: Stratification for Priority Survey
[What to start with]

[Procedure 1]   Confirmation of Occurrence Status

Figure 4.9 shows a Pareto chart of defective assembly for priority investigation to improve the quality defect items of product B. From Figure 4.9 it was found that 90% of the metal fittings were defective and that it is essential to improve these defects in order to achieve the target. Figure 4.10 shows the dimensional measurement part of the metal fitting position.

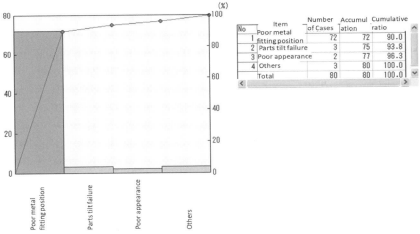

No	Item	Number of Cases	Accumulation	Cumulative ratio
1	Poor metal fitting position	72	72	90.0
2	Parts tilt failure	3	75	93.8
3	Poor appearance	2	77	96.3
4	Others	3	80	100.0
	Total	80	80	100.0

Figure 4.9 Pareto Chart of Poor Assembly

[Procedure 2]   Setting Goals

As a goal, the metal fitting position was set to zero in order to aim for extraordinary quality (1/10 of the defect rate of product B).

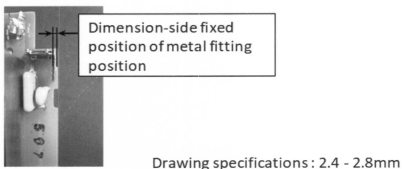

Dimension-side fixed position of metal fitting position

Drawing specifications : 2.4 - 2.8mm

Figure 4.10 Metal Fitting Position, Dimension Measurement Site

[Procedure 3] **Preventing the Outflow of Defective Products to Customers.**

To prevent outflow to customers, we carry out a 100% visual inspection.

Step 3: Quantification of Defect Characteristics
[Reliability confirmation of quality characteristic evaluation]

[Procedure 1] **Measure the Quality Characteristics of Non-Defective and Defective Products.**

Figure 4.11 shows a histogram of the distribution of survey lot A which is a mixture of non-defective and defective products. It was confirmed that reliable judgement of non-defective products and defective products is possible.

[Procedure 3] **Implementation of Gauge R & R (GR & R)**

When the reliability of the measurement by calipers was evaluated by GR & R, the GR & R value was statistically proved to be 10% or less.

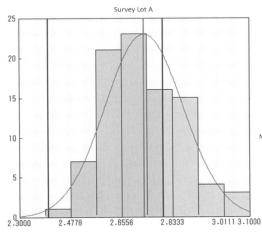

Variable number	5
Number of data	90
Minimum value	2.460
Maximum value	3.050
Average value	2.7342
Standard deviation	0.13575
Distortion	0.200
Sharpness	-0.549
Upper limit standard value	2.800
Lower limit standard value	2.400
Cp	0.491
Cpk	0.162
Number of non-standard data	30

Figure 4.11 Histogram of Metal Fitting Position Dimensional Distribution of Survey Lot A

Step 4: Stratification for Factor Investigation (I) [Phenomenon] [Search for factors from the result system]

[Procedure 1] Is There a Trend Among Each Product?

Figure 4.12 shows a Pareto chart for comparing the occurrence of defects for product B and similar product K. From Figure 4.12 it turned out that poor metal fitting position doesn't occur in other products and is a defect peculiar to product B.

[Procedure 2] Is There Any Variation Between Manufacturing Processes?

It turned out that poor metal fitting position occurred during the assembly process (manual work).

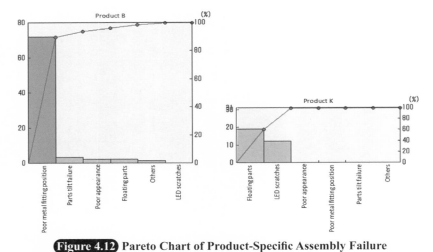

Figure 4.12 Pareto Chart of Product-Specific Assembly Failure

Step 5: Stratification for Factor Investigation (II) [Details] [Search for factors from the causative system]

[Procedure 1] Product function check

When the positional relationship from the state where the metal fittings are attached in Figure 4.10 to the state before the metal fittings are attached is checked, it can be seen that the position of the hole from one side near the place of occurrence is directly linked to the metal fitting position failure as shown in Figure 4.13.

It was also found that the supplier of the printed circuit board for similar product K was different from other products. It is the assembly process that causes the metal fittings to be misaligned but it turned out that the accuracy of the purchased parts is also likely to be related.

Also when this purchasing parts manufacturer was asked if the boards were produced on multiple lines, they said that after three boards are simultaneously molded, they are cut into three. Figure 4.14 shows the printed circuit board be-

Measurement

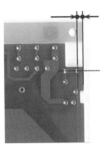

Figure 4.13 Purchased Parts Printed Circuit Board Metal Fitting Mounting Part Diagram

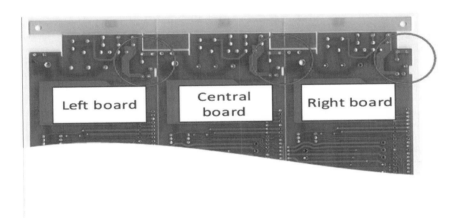

Figure 4.14 Purchased Parts Printed Circuit Board Diagram (Before Cutting)

Success Stories Using SOUTH-FLOW

Chapter 4

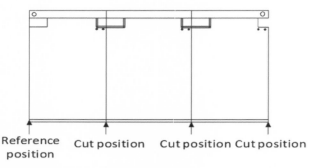

Reference position   Cut position   Cut position   Cut position

Figure 4.15 **Purchased Parts Printed Circuit Board Standard Cut Diagram**

fore cutting. The circled positions indicate the parts related to the metal fitting position failure shown in Figure 4.13.

[Procedure 2]   **Confirmation of Equipment Functions**

As a condition of the equipment, it was found that the printed circuit board was manufactured as a set of three and the cutting standard was one side of the left side board as shown in Fig. 4.15. Depending on the workmanship of the three cut dimensions, it is expected that the misalignment of metal fittings occurs.

[Procedure 4]

Figure 4.16 shows the results of investigating the dimensions of Figure 4.13 for each of the three types of boards in a set of three boards in a histogram. From Figure 4.16 there were no defective products on the left board but defects were confirmed on the center board and right board. In particular, the difference of a higher defect rate as it goes to the right was found.

Furthermore, when one checked the width dimensions of each of the three types of boards, they found that although they were all large, there were no particular problems in terms of drawing specifications. Figure 4.17 shows the distribution of the three width dimensions in a histogram.

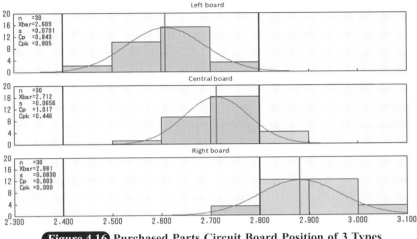

Figure 4.16 Purchased Parts Circuit Board Position of 3 Types of Cut Histogram

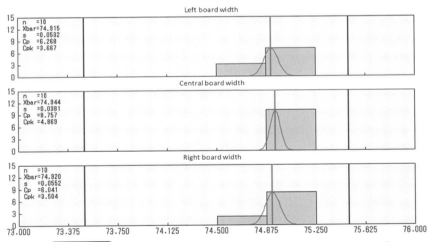

Figure 4.17 Purchased Parts Circuit Board Width Dimension of 3 Types of Cut Histogram

Step 6: Organization of Factors
[Organize factors with the strongest members using facts as hints]

[Procedure 1] Arrangement of the Relationship Between Defect Characteristics and Factors

In Step 6, key people in the company were asked to gather and pay attention to the variation in cut dimensions in the outer shape cutting process of board manufacturing, not in the assembly process based on the factors, intuition, and experience discovered in Step 5 as shown in Figure 4.18. Then the factors were sorted out.

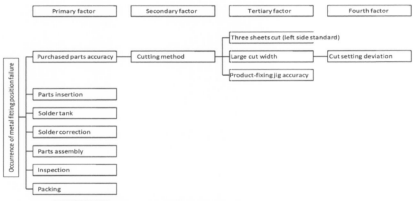

Figure 4.18 Causes of Mold Position Defects System Diagram

Steps 7 and 8: Experiments for Factor Analysis (I), (II)

The factors this time were the inadequate set value of the cut dimension of a Chinese purchasing parts manufacturer and an inadequate fixing jig for the product at the time of cutting. So the experiment was deemed unnecessary and omitted.

Step 9: Implementation of Measures
[Introduction of optimal conditions for mass production]

[Procedure 2] Implementation of Mass Production Measures

As measures incorporated into mass production, review of cut dimension settings at the Chinese purchasing parts manufacturer and improvement of the product fixing jig at the time of cutting were conducted.

Step 10: Effect Confirmation
[Confirmation of effects after introduction of mass production measures]

[Procedure 1] Confirmation of the Effect of Improvement

Figure 4.19 shows the position and dimension distribution of metal fittings using the improved parts of the purchased parts manufacturer in China. As can be seen by comparing the distribution of component dimensions in Figure 4.16, there were no defective products on the left board, center board, right board all boards, and we were able to clear the goal. The process capability also improved from 0.81 to 1.15 on the left board, from 0.45 to 1.28 on the center board and from 0.00 to 1.05 on the right board.

[Procedure 3] Grasping the Results and Confirming the Spillover Effect

The improvement leader (himself) and the Chinese support members were able to experience successful quality improvement using SOUTH-FLOW and gained confidence to take on the challenge of the next improvement theme.

Step 11: Standardization
[Succession of improvement measures to the next generation]

Omitted from print.

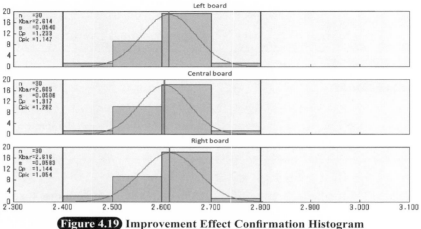

Figure 4.19 Improvement Effect Confirmation Histogram

~ **Impressions About Using SOUTH-FLOW** ~

With SOUTH-FLOW, we were able to solve the problem by ordering and organizing the problem-solving procedure.

Initially, it was an improvement scenario in which I thought it was poor work done by the workers, but with SOUTH-FLOW the worker-related investigation was the final step and improvement was possible without making a detour.

(Interview with Mr. S, Improvement Leader of Company C)

4.3 Development of New F Substrate Adhesive X at Resin Manufacturing Company A

The improvement leader is Mr. S from the F Circuit Research Department. Company A worked on the development of products that had not been solved so far by applying SOUTH-FLOW.

The manufacturing process is shown in Figure 4.20.

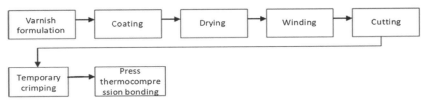

Figure 4.20 **Schematic Diagram of the Manufacturing Process**

The material manufacturing process is (1) varnish formulation → (2) coating (coating machine) → (3) drying (coating machine) → (4) winding (coating machine). The usage is (5) cutting → (6) temporary crimping → (7) press thermocompression bonding.

> ## Step 1: Problem-Solving Solution
> ## [Scenario selections for pursuing the root cause]

[Procedure 1] **Solution Theme Selection**

Company A is developing a new F circuit board. They have adhesives that are currently being produced and used but they do not meet the requirements of new F circuits. They have other divisions that handle adhesives but are not satisfied with similar products. Therefore, they decided to aim to complete a new F circuit board by developing a new adhesive X in their department. The main direction of development is the improvement of the conventional type with the aim to complete it by adding new technology to it.

Success Stories Using SOUTH-FLOW

[Procedure 2] **Registration of Theme Name**

The theme name is a development of adhesive X for new F substrate.

> ## Step 2: Stratification for Priority Survey
> ## [What to Start With]

[Procedure 1] **Confirmation of Occurrence Status**

Figure 4.21 shows our current product A and new in-house required standards. From Figure 4.21 it can be seen that both adhesion and resin flow do not meet the standards. Figure 4.22 shows the adhesion evaluation test to the substrate and Figure 4.23 shows the resin flow evaluation. In particular, median (x) was also used to improve the reliability of resin flow evaluation.

Understanding the current situation	Adhesion to substrate (N/mm)	Resin flow (mm)
New in-house required standard	0.85~1.50	0~0.50
Current product A	0.22	1.21

Figure 4.21 Comparison Between New In-House Required Standards and Current Products

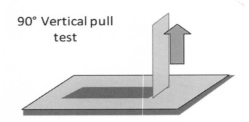

90° Vertical pull test

Figure 4.22 Schematic Diagram of Adhesion Evaluation Test

Resin flow evaluation

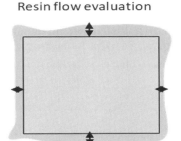

Figure 4.23 Resin Flow Evaluation Chart

[Procedure 2]   Goal Setting

The goal was to develop adhesive X which meets the new in-house required standard in Figure 4.21. The sub-theme is the development of adhesion-improved products that meet resin flow standards.

Step 3: Quantification of Defect Characteristics [Reliability confirmation of quality characteristic evaluation]

[Procedure 3]   Implementation of Gauge R & R (GR & R)

First, it was confirmed by GR & R that a test method that keeps the value within 10% was established.

Step 4: Stratification for Factor Investigation (I) [Phenomenon] [Finding factors from the result system]

[Procedure 1]   Is There a Trend Among Each Product?

Figure 4.24 shows a comparison of quality characteristics of similar products. From Figure 4.24, none of the products met the two quality characteristics at the same time. However, it was also found that only similar product B satisfied the resin flow standard.

Understanding the current situation	Adhesion to substrate (N/mm)	Resin flow (mm)
New in-house required standard	0.85~1.50	0~0.50
Current product A	0.22	1.21
Similar product B	0.04	0.5
Similar product C	0.64	1.24

Figure 4.24 Comparison of quality characteristics of similar products

Step 5: Stratification for Factor Investigation (II) [Details] [Search for factors from the causative system]

[Procedure 1] Product Function Confirmation

Depending on the composition of the adhesive, it is known that there are differences in quality characteristic values. However, the degree of influence on which quality characteristics, the entanglement of influence between components, etc. isn't clear.

Step 6: Organization of Factors
[Organize factors with the strongest members using facts as hints]

[Procedure 1] Arrangement of the Relationship Between Defect Characteristics and Factors

While investigating the components of each of the three types of adhesives with the cooperation of the analysis group, the relevant key people were asked to gather and sort out the factors that can and cannot be managed in-house regarding adhesion as shown in Figure 4.25.

Steps 7 and 8: Experiment for Factor Analysis (I), (II)

[Procedure 1] **Experiment by Allocating Contributing Factors to Three Levels**

Together with the key people, possible factors were narrowed down to five. In addition, the members suggested that there were three possible entanglements, and said, "I don't want to waste the research so far, so I want to experiment at three levels." With the superior of the key people acknowledging their request, it was decided to carry out the L27 orthogonal experiment twice and to analyze the factors. Figure 4.26 shows the factors given to the key people in a matrix.

Figure 4.27 shows the analysis of variance tables as the result of the L27 orthogonal array experiment on adhesion.

It was found that factors A, B, C and AC have a large effect on adhesion.

Figure 4.28 shows the degree of influence of factors A, B, C and AC on adhesion in a factor effect plot diagram.

From Figure 4.28 it is possible to see the intertwining of factor A and factor C and it seems that the key people were also thinking about new discoveries.

Chapter
4

Success Stories Using SOUTH-FLOW

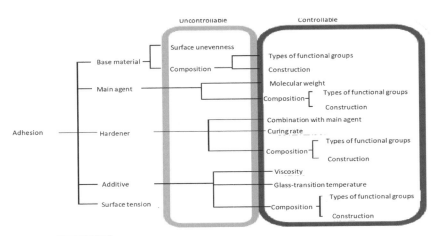

Figure 4.25 Systematic Diagram of Factors Related to Adhesion

Figure 4.29 shows the combinations that were able to meet the resin flow and adhesion standards at the same time. Among them it was found that selecting level 2 for factor A, selecting level 3 for factor B and setting factor C to level 2 shows the optimal value (the evaluation result for resin flow is omitted).

Factor	Main Effect	Level		
		1	2	3
Main agent	A	a	b	c
Hardener	B	d	e	f
Additive	C	g	i	j
	F	h	n	None
Flame retardants	D	k	l	m

Figure 4.26 Factors and three levels matrix

Factor	Sum of Squares	Flexibility	Variance	Dispersion Ratio	Test	P-value (upper side)
A	1.95	2	0.975	288.638	**	0
B	0.217	2	0.109	32.156	**	0
C	0.823	2	0.411	121.763	**	0
D	0.007	2	0.003	0.998		0.38
F	0.021	2	0.011	3.173		0.056
AB	0.015	4	0.004	1.082		0.382
AC	0.057	4	0.014	4.197	**	0.008
BC	0.019	4	0.005	1.42		0.251
Measurement Error	0.105	31	0.003			
Total	3.213	53				

Figure 4.27 L27 Orthogonal Experiment Results on Adhesion (Analysis of Variance Table)

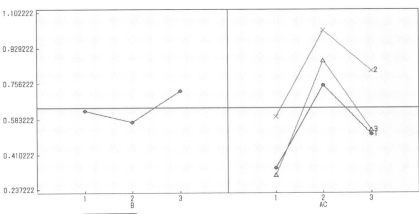

Figure 4.28 Main Effects Plot Related to Adhesion

Factor	Estimated Value	Confidence interval	
ABC		Lower limit	Upper limit
232	1.10	1.05	1.16
212	1.00	0.95	1.06
233	0.96	0.90	1.01
222	0.95	0.90	1.00

Figure 4.29 Table of Combinations that Satisfy Both Resin Flow and Adhesion

Step 10: Effect Confirmation
[Confirmation of effects after introduction of mass production measures]

[Procedure 1] Confirmation of Improvement Effect

Confirmation of other quality characteristics for mass production are also under way. This time, the discovery of a combination that satisfies the most important quality characteristics was a great yield for the department. Also, in the opinion of the key people, this discovery is likely to be patentable.

Step 11: Standardization
[Succession of improvement measures to the next generation]

Omitted from print.

∼ Impressions About Using SOUTH-FLOW ∼

The detailed flow chart makes it clear what to do next and there is no going back and forth like "Oh, I forgot that" or "Oh, I have to do this" which is common in business; since it's possible to go in one direction time is saved.

In particular, I like the fact that we are paying attention to the connections between people, such as "incorporating key people," etc.

(Interview with Mr. S, Improvement Leader of company A)

4.4 What We learned From Success Stories in Three Different Industries

Figure 4.30 shows the main SOUTH-FLOW processes and QC methods that have been effective in improvement in the three successful cases provided this time.

Looking at the process of searching for the root cause, it is possible to see that the step topics that worked for improvement are different in each case.

On the other hand, looking at the process of forming the strongest team, all three cases have two common processes. The first thing that they have in common is that they incorporated key people needed for improvement. The second thing that they have in common is the improvement of their ability to solve problems. It can be said that this led to the discovery of the true cause that had not been noticed until now, the result of improving the eyes to see the site due to the results of the factor investigation visualized by the improvement leaders and the key people who gathered. Therefore, by discovering the root cause in all three cases, appropriate measures were planned and implemented leading to success.

Finally, comparing the main QC methods used, GR & R is used when confirming the reliability of measurement because all the improvement cases had measurable quality characteristics. In addition, QC Seven Tools such as histograms and Pareto charts, and the new QC Seven tools such as system diagrams and matrices are used. This result also shows that basic QC methods are useful for visualizing on-site information in activities to eliminate chronic defects.

No.	Success Story Theme Name Industry / Department of Improvement Leader	The Process of Searching for the Main Root Causes of Improvement	The Process of Forming the Main Strongest Team That Worked for Improvement	Main QC methods Used for Improvement, etc. (*): Introduced in Chapter 3
1	· Improvement of measurement defect rate of subdivided packaging. · Pharmaceutical manufacturing industry N company / manufacturing department	Step 3 Measurement reliability Step 6 Factor organization	III. Incorporation of key people IV. Improvement of solution	GR & R (*) Control chart, histogram, Ishikawa diagram
2	Improved defective rate of printed circuit board product B IT company C / Quality Assurance Department	Step 2 Priority survey Step 4 Result-based factor investigation Step 5 Investigation of causative factors	III. Incorporation of key people IV. Improvement of solution	GR & R (*) Pareto chart, histogram, system diagram
3	Development of adhesive X and improvement of adhesiveness and resin flow Company A / Research Division	Step 3 Measurement reliability Step 4 Result-based factor investigation Step 8 Factor analysis experiment	III. Incorporation of key people IV. Improvement of solution	GR & R (*) Matrix, system diagram, Design of experiments(*)

Figure 4.30 Features of Three Cases of SOUTH-FLOW Utilization

Appendix 1 Support for Improvement Activities Using Software Equipped with SOUTH-FLOW

~ StatWorks / QCAS V4.0 (Chronic defect elimination version) ~

1. SOUTH-FLOW-Equipped Software, Chronic Defects Elimination Version

JUSE-StatWorks / QCAS (Chronic Defect Elimination Edition) is a problem-solving tool equipped with SOUTH-FLOW. It is an optimal system in which procedures, analysis tools, progress management, document management, etc. can be used to efficiently support SOUTH-FLOW. In particular, various tools are integrated to speed up problem solving of chronic defects, such as systematization of procedures and hints, incorporation of analysis tools, progress management (analysis review list), automatic document management, etc.

Usually, regarding the solution of chronic defects, 1) we do not know how to solve chronic defects, 2) we do not have analytical tools and other tools, 3) there are no human resources who can solve problems, 4) there is not enough management who can support it. In addition, there are problems such as the statistical methods and analysis tools learned in the training cannot be used in the field and the documents after problem solving become complicated and cannot be managed. The use of analytical tools with SOUTH-FLOW provides one way to solve these problems.

The features of the chronic defect elimination version equipped with SOUTH-FLOW are as follows.

《Features》

1. Systematization of the problem-solving procedure for chronic defects into 11 problem-solving steps. The know-how cultivated through improvement activities in the manufacturing industry is concentrated in SOUTH-FLOW and it is equipped with scenarios that are easy for practitioners to understand.

2. All the analytical tools used in each step are included in the set, making

it easy to use. There are many useful functions for practitioners such as progress management tools, hints according to situation, forms, etc.

3. The "Rolling paper" function makes it easy to manage and utilize documents such as reports, presentations, etc.

4. StatWorks / V4.0 (or QCAS), which has a proven track record in quality control in Japan, is fully packed and all the analysis methods included can be used.

The main steps to take advantage of the Chronic Defect Elimination Edition are below.

(1) Registration of activity theme...click on the theme button on the main screen.

(2) Collect survey data according to procedure... easy-to-understand in the action panel.

(3) Data entry and analysis...use of easy-to-understand input formats and reliable analytical tools.

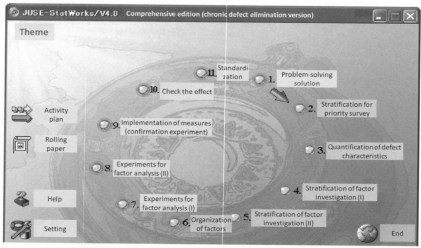

Figure a Main Screen for Eliminating Chronic Defects

(4) Organize analysis results step by step...use of the Rolling paper function.

(5) Preparation of activity report...created in standard format.

Figure a shows the main menu of the StatWorks / V4.0 Chronic Defect Elimination Edition with SOUTH-FLOW. You can click here to proceed with the 11 steps of SOUTH-FLOW.

2. Action Panel and Rolling paper Functions

Figure b is a screen called "Action Panel" that displays the procedures and points of recommendation in each step, recommended statistical methods, data files, etc.

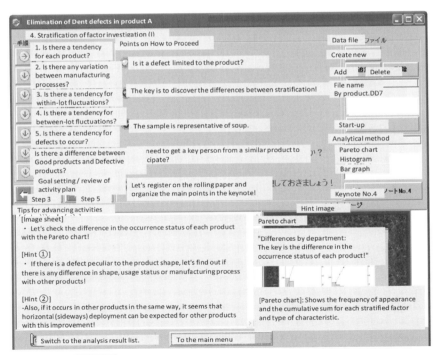

Figure b Action panel and analysis result review window

The "Analysis Result Review Window" is displayed at the bottom of the screen. Since the screens and images saved during the analysis are displayed here, you can take the next action while checking the analysis results that the user has carried out so far.

In addition, the Rolling paper lists the StatWorks screen output files saved during the analysis and the images (formats such as png and jpg) used there for each step, etc. as shown on the left side of Figure c. In addition, you can check the analysis results and easily paste the required screens into Excel or Power-Point such as the improvement report (standard equipment as a keynote) in Figure 2.47 in Chapter 2.

3. Progress Confirmation List

The progress review list is a function that allows you to see at a glance how

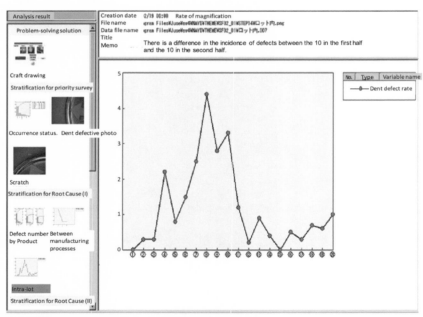

Figure c Rolling paper function that can collect and edit analysis results and images

far the analysis has been performed during the improvement activities (Figure d). You can easily see what you have done in each step or procedure.

If you look at the checkpoints, you can see a list of how far the data analysis has progressed, which procedure you have not performed, etc. and you can directly and quickly display the steps you want to check. In carrying out improvement activities, it will be a common base for improvement practitioners and managers to evaluate and confirm progress.

4. Improvement Bulk Filing System

Figure e shows the data file copy operation screen for document management in improvement activities.

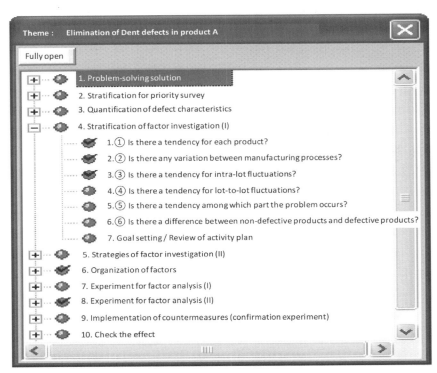

Figure d Progress review list

Appendix 1 Support for Improvement Activities Using Software Equipped
with SOUTH-FLOW

Elimination of defective Dent in product A

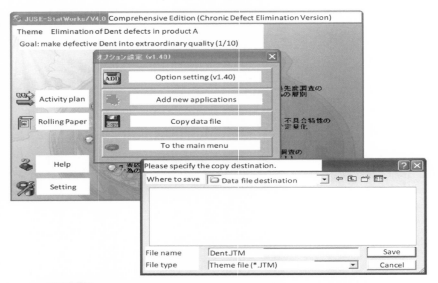

Figure e Bulk Document Management in Improvement Activities

Appendix 2 List of Analysis Methods That Can be Used in the StatWorks V4.0 Chronic Defect Elimination Version

The information distributed on the personal computer or server after completion of the improvement activities are completed and is put together in an arbitrary folder while maintaining the connection between the system and the document.

Therefore, during or after the completion of the improvement activities, it is collectively stored at the site and by sending the files to the quality control office, the contents are verified and advice is given. There are several ways to send the file, such as attaching it to an e-mail after compression, copying it to flash memory, etc.

Appendix 2 List of Analysis Methods Available in StatWorks V4.0 Chronic Defect Elimination Edition

Method				Stat-Works	QCAS
基本解析 Basic analysis	統計量 / 相関係数 Statistics / Correlation coefficients			●	●
	度数表 / 多変量クロス表 Frequency table / Multiple cross table			●	●
	多変量連関図 MA chart (Multivariate association chart)			●	●
	モニタリング Quick view			●	●
	一般グラフ General graph	折れ線グラフ Line graph		●	●
		棒グラフ Bar graph		●	●
		帯グラフ Band graph		●	●
		円グラフ Pie chart		●	●
		レーダーチャート Radar chart		●	●
		XY 折れ線グラフ XY line graph		●	●

Appendix 2 List of Analysis Methods That Can be Used in the StatWorks V4.0 Chronic
Defect Elimination Version

Category	Method	Sub-method		
QC七つ道具 QC Seven Tools	特性要因図 Ishikawa diagram (Cause-and-effect diagram)		●	●
	パレート図 Pareto chart		●	●
	ヒストグラム Histogram		●	●
	管理図 Control chart		●	●
	管理図（新 JIS）Control chart (new JIS)		●	●
	散布図 Scatter diagram		●	●
	グラフ Graph	折れ線グラフ Line graph	●	●
		棒グラフ Bar graph	●	●
		帯グラフ Band graph	●	●
		円グラフ Pie chart	●	●
		レーダーチャート Radar chart	●	●
		ワイブル確率紙 Weibull probability paper	●	●
		XY 折れ線グラフ XY line graph	●	●
新QC七つ道具 New QC Seven Tools	親和図 Affinity diagram		●	●
	連関図 Relational diagram		●	●
	系統図 System diagram		●	●
	マトリックス図 Matrix diagram		●	●
	マトリックスデータ解析 Matrix data analysis		●	
	PDPC PDPC		●	●
	活動計画表（ガントチャート）Activity plan (Gantt chart)		●	●
工程分析 Process analysis	SPC（工程性能分析）SPC (Process Capability Analysis)		●	●
	MSA（ゲージ R&R 他）Measurement Systems Analysis (Gauge R & R, etc.)		●	●
	FMEA Failure Mode and Effects Analysis		●	●
	品質機能展開（QFD）Quality Function Deployment (QFD)		●	●
	累積和管理図（CUSUM）Cumulative Sum Control Chart (CUSUM)		●	●
	多変量管理図 Multivariate control chart		●	●

		食品衛生管理図 Food hygiene control chart	●	●
実験計画法 Experimental design	要因配置実験 Factor placement experiment	要因配置実験の計画 Factor placement experiment plan	●	●
		一元配置分散分析 ANOVA: One-way layouts	●	●
		二元配置分散分析 ANOVA: Two-way layouts	●	●
		多元配置分散分析 ANOVA: Multi-way layouts	●	●
	直交表実験 Orthogonal array experiment	直交表実験のための計画 Plan for orthogonal array experiments	●	●
		直交配列表 Orthogonal sequence list	●	●
	応答曲面法 Response surface methodology	応答曲面法のための計画 Planning for Response surface	●	●
		1特性の最適化 Optimization of Single response	●	●
		多特性の最適化 Optimization of Multiple response	●	●
	品質設計（タグチメソッド） Quality design (Taguchi method)	パラメータ設計のための計画 Setting of Robust Parameter design	●	●
		パラメータ設計 Parameter design	●	●
		許容差解析 Tolerance analysis	●	●
	効果プロット Effect plot		●	●
	マルチバリチャート Multi-vari chart		●	●

Appendix 2 List of Analysis Methods That Can be Used in the StatWorks V4.0 Chronic Defect Elimination Version

回帰分析 Regression analysis	単回帰分析 Simple regression analysis	●	●
	重回帰分析 Multiple regression analysis	●	●
	直交多項式回帰分析 Orthogonal polynomial regression analysis	●	●
	ロジスティック回帰分析 Logistic regression analysis	●	●
	重みつき回帰分析 Weighted regression analysis	●	
多変量解析 Multivariate analysis	主成分分析 Principal component analysis	●	
	重回帰分析・数量化Ⅰ類 Multiple regression analysis / Quantification method I	●	
	判別分析・数量化Ⅱ類 Discriminant analysis / Quantification method II	●	
	数量化Ⅲ類 Quantification method III	●	
	ロジスティック回帰分析 Logistic regression analysis	●	
	多段層別分析（AID） Multi-tiered analysis (AID)	●	
調査分析 Survey analysis	集計表解析 Schedule analysis	●	
	SD法（SDプロファイル） SD method (SD profile)	●	
	コンジョイント分析 Conjoint analysis	●	
	因子分析 Factor analysis	●	
	非階層的クラスター分析（k-means法クラスター分析） Non-hierarchical cluster analysis (k-means cluster analysis)	●	
	階層的クラスター分析 Hierarchical cluster analysis	●	
時系列解析 Time series analysis	時系列グラフ Time series graph	●	
	循環図 Circulation diagram	●	
	回帰による要因分解 Factor decomposition by regression	●	
	ARIMAモデル ARIMA model	●	

信頼性解析 Reliability analysis	グラフによる観察 Observation by graph	解析線図 Analysis diagram	●	
		状態線図 State diagram	●	
		信頼度・故障率曲線 Reliability / Bath-tub curve	●	
	確率紙 Probability paper	確率紙（ワイブル確率ほか） Probability paper (Weibull probability, etc.)	●	●
		確率プロット Probability plot	●	
	確率密度曲線 Probability density curve		●	
	分布型の検定・推定 Distribution type test / estimation	最尤推定 Maximum likelihood estimation	●	
		カイ二乗検定 Chi-square test K-S 検定 K-S test	●	
	時系列グラフ Time series graph		●	
	市場データの解析 Analysis of market data	CHM CHM	●	
		クロス表変換 Cross table conversion	●	
		ワイブル型累積ハザード紙 Weibull type cumulative hazard paper	●	
	日付変換 Date conversion		●	
検定・推定 Test / estimation	計数値の検定・推定 Test / estimation of count value	母不良率 Nonconforming ratio in population	●	●
		母不良率の差 Difference of Nonconforming ratio in two population	●	●
		母欠点数 Number of defects in population	●	●
		母欠点数の差 Difference of defects in population	●	●
		M × n 分割表 Mn contingency table	●	●

Appendix 2 List of Analysis Methods That Can be Used in the StatWorks V4.0 Chronic Defect Elimination Version

		計量値の検定・推定	母分散 Population variance	●	●
			2つの母分散の比 Ratio of two population variances	●	●
			母平均 Population mean	●	●
			2つの母平均の差 Difference between two population means	●	●
			データに対応がある場合の母平均の差 Difference for the two population means for the paired data	●	●
			3つ以上の母分散の一様性 Uniformity of three or more population variances	●	●
		ノンパラメトリック検定	カイ二乗検定 Chi^2 Test	●	●
			ウイルコクソン順位和検定 Wilcoxon rank sum test	●	●
			MOOD 検定 MOOD test	●	●
			クラスカル・ウォリス検定 Kruskal-Wallis test	●	●
			ウイルコクソン符号付順位和検定 Wilcoxon rank sum test	●	●
			フリードマン検定 Friedman Test	●	●
	検出力 Detectability	検出力とサンプルサイズ	母分散 Population variance	●	●
			2つの母分散の比 Ratio of two population variances	●	●
			母平均 Population mean	●	●
			2つの母平均の差 Difference between two population means	●	●
	確率値 Probability value		データに対応がある場合の母平均の差 Difference for the two population means for the paired data	●	●
		連結分布 Continuous distribution		●	●
		離散分布 Discrete distribution		●	●
プロット表示ツール Plot display tool				●	●

(Note) The shaded method is explained in Chapter 3 of this book.

Appendix 3 Download from homepage

To obtain JUSE-StatWorks V4.0 "Chronic Defect Elimination Edition" (30-day trial limited edition), you can download the trial version from the following Nikkagiken Union of Japanese Scientists and Engineers website "JUSE-Communication on WEB."

http://www.i-juse.co.jp/statistics/index.html

►References

(1) Nichogi Hideo, *Can you say that you are a manager without knowing this management method?* (Chukei publish, 1995)

(2) Shibuya Shozo, *Interesting and instructive psychology trivia* (NIPPON JITSUGYO PUBLISHING, 1998)

(3) Iwasaki Hideo, "Problem Solving Method (Chapter 2),", *Quality Control Seminar Basic Course Text* Ed. (Japan Science and Technology Federation-nUnion of Japanese Scientists and Engineers, 2001)

Kenji Minami 南 賢治 (みなみ けんじ)

Born in 1963
1987 Graduated from Shibaura Institute of Technology, Faculty of Engineering and joined NHK Spring Co., Ltd.
He was assigned to quality control, production engineering, SQC promotion, and quality education for automotive, furniture, hard disk drives, and home appliances. He is currently assigned to the Chief of Quality Assurance Department.

Kiyoshi Katayama 片山 清志 (かたやま きよし)

Born in 1955
Graduated from Tokyo Institute of Technology with a degree in Engineering in 1979.
Withdrew from the doctoral program of Graduate School of Science and Engineering, Tokyo Institute of Technology and joined the Institute of Japanese Union of Scientists & Engineers in 1986.
He has been assigned to development, support, and corporate training of various statistical packages and QMS, such as JUSE - QCAS, RAS1,
StatWorks, and Sutatto! for 35 years. He is now the General Manager of the Mathematical Department.

SOUTH-FLOW - The Original and Effective Method of Quality Improvement Scenario Specializing in Eliminating Chronic Defects
February 25, 2005 1st printing
February 12, 2008, 2nd printing
Author Kenji Minami Kiyoshi Katayama
Publisher: Hiroyoshi Taniguchi
Published by JUSE Publishing Co.
5-4-2 Sendagaya, Shibuya-ku, Tokyo 151-0051, Japan
Phone: Publications 03-5379-1244
Sales department 03-3379-1238~9
Transfer account : Tokyo 00170-1-7309
Printed in Japan
© Kenji Minami, Kiyoshi Katayama 2005 ISBN97848171-91328
(This publication may not be reproduced in whole or in part, with the exception of copyright law.)
URL http://www.juse-p.co.jp/

Yes, you can! How to eliminate chronic defects

SOUTH-FLOW - The Original and Effective Method of Quality
Improvement Scenario Specializing in Eliminating Chronic
Defects

2022年9月11日　初版第1刷発行

著　者	南　賢治	
	片山清志	
訳　者	南　賢治	
発行者	中田典昭	
発行所	東京図書出版	
発行発売	株式会社 リフレ出版	
	〒113-0021　東京都文京区本駒込3-10-4	
	電話 (03)3823-9171　FAX 0120-41-8080	
印　刷	株式会社 ブレイン	

© Kenji Minami, Kiyoshi Katayama
ISBN978-4-86641-487-4 C3050
Printed in Japan 2022

本書のコピー、スキャン、デジタル化等の無断複製は著作権法上
での例外を除き禁じられています。本書を代行業者等の第三者に
依頼してスキャンやデジタル化することは、たとえ個人や家庭内
での利用であっても著作権法上認められておりません。

落丁・乱丁はお取替えいたします。
ご意見、ご感想をお寄せ下さい。